2nd International Conference on Artificial Intelligence, Automation and Control Technologies (AIACT 2018)

Journal of Physics: Conference Series
Volume 1061

Osaka, Japan
26 – 29 April 2018

ISBN: 978-1-5108-6828-1
ISSN: 1742-6588

Printed from e-media with permission by:

Curran Associates, Inc.
57 Morehouse Lane
Red Hook, NY 12571

Some format issues inherent in the e-media version may also appear in this print version.

Printed by Curran Associates, Inc. (2018)

For permission requests, please contact the Institute of Physics
at the address below.

Institute of Physics
Dirac House, Temple Back
Bristol BS1 6BE UK

Phone: 44 1 17 929 7481
Fax: 44 1 17 920 0979

techtracking@iop.org

Additional copies of this publication are available from:

Curran Associates, Inc.
57 Morehouse Lane
Red Hook, NY 12571 USA
Phone: 845-758-0400
Fax: 845-758-2633
Email: curran@proceedings.com
Web: www.proceedings.com

Table of contents

Volume 1061

2018 2nd International Conference on Artificial Intelligence, Automation and Control Technologies (AIACT 2018)

26–29 April 2018, Osaka, Japan

Accepted papers received: 5 July 2018
Published online: 19 July 2018

Preface

Papers

Artificial Intelligence and Automation

IOP Conf. Series: Journal of Physics: Conf. Series **1061** (2018) 011001 doi:10.1088/1742-6596/1061/1/011001

PREFACE

2018 2nd International Conference on Artificial Intelligence, Automation and Control Technologies (AIACT 2018) is an annual conference which offers a place for participants to present and to discuss their innovative recent and ongoing research and theoretical work and/or their applications. The first AIACT was held in China University of Geosciences, Wuhan, China in 2017. This year is the second AIACT annual conference, which was held in Osaka, Japan, during April 26-29, 2018. AIACT 2018 is organized by Hong Kong Society of Mechanical Engineers, sponsored by York University. The conference provides a useful and wide platform for researchers and engineers to present their latest research findings and exchange their ideas in a broad range of Artificial Intelligence, Automation and Control Technologies topics. The participants of the conference were from different regions of the world, with background of either academia or industry.

There were 87 submissions including 69 papers and 18 abstracts. After rigorous peer review, 27 papers and 6 abstracts were accepted. This proceeding covers a wide range of topics in Artificial Intelligence, Automation and Control Technologies, including Adaptive Control, Control Theory and Application, Control theory & applications optoelectronics, Automation Engineering, Automated Guided Vehicles, etc.

We would like to thank all of those who supported AIACT 2018. Particularly, we would like to thank the organizing committee for their invaluable advices in the organization and the program committee for the peer review of the papers.

We sincerely hope that AIACT will be an excellent forum for discussions that will lead to fruitful research studies and other new scientific ideas and engineering products that will be beneficial to the society and the industry.

Prof. Dan Zhang, York University, Canada

COMMITTEES

Conference Chair

Prof.Dan Zhang, York University, Canada

Program Committee

Prof.Jiann-Shiou Yang, University of Minnesota, USA

Prof.Songyi Dian, Sichuan University, China

Local Chair

Prof.Weiwei Wan, Osaka University, Japan

International Technical Committee

Prof.Wei Dong, Harbin Institute of Technology, China

Prof.Samir N. Mekid, King Fahd University of Petroleum & Minerals, Saudi Arabia

Prof. Johan Potgieter, Massey University, New Zealand

Prof. Chan Chun Chung Keith, The Hong Kong Polytechnic University, Hong Kong SAR

Prof. Peter Weiliang Xu, University of Auckland, New Zealand

Prof. Kefu Liu, Lakehead University, Canada

Prof. Huang Jie, The Chinese University of Hong Kong, Hong Kong SAR

Prof. Shih-Chi Chen, The Chinese University of Hong Kong, Hong Kong SAR

Prof.Ross McAree, University of Queensland, Australia

Dr. Wei Dong, Harbin Institute of Technology, China

Dr. Zhao Tao, Sichuan University, China

Dr. Steven Dirven, Massey University, New Zealand

Dr. Li Wen, Beihang University, China

IOP Conf. Series: Journal of Physics: Conf. Series **1061** (2018) 011003 doi:10.1088/1742-6596/1061/1/011003

Peer review statement

All papers published in this volume of *Journal of Physics: Conference Series* have been peer reviewed through processes administered by the proceedings Editors. Reviews were conducted by expert referees to the professional and scientific standards expected of a proceedings journal published by IOP Publishing.

Collision Hazard Identification of Unmanned Vessels in Inner River Based on Particle Swarm Parameter Optimization Support Vector Machine

Wenli Ma, Jie Yang*, Qingnian Zhang, ZhiqiangGuo and Rong Hu

Key Laboratory of Fiber Optic Sensing Technology and Information Processing, Ministry of Education, Wuhan University of Technology,Wuhan, China

* 794393254@qq.com

Abstract. In this thesis, Collision hazard identification of unmanned vessels in inner river is studied. The electronic chart, radar and AIS system of ship assembly can provide the data that affect the normal operation of unmanned ships. The fuzzy comprehensive evaluation method is used to evaluate the collision of unmanned ships Levels of danger. The particle swarm optimization algorithm is used to support Vector machines classify the ship's collision hazard classification. The data is based on data from 112 survey reports of collision between two ships issued by various provinces and local maritime authorities in China during 2004-2015 as a simulation sample. The collision risk of unmanned ships is divided into three levels, and Proposed a different level of unmanned ships coping methods. By comparing experimental results with ordinary support vector machine, particle swarm optimization, genetic algorithm and grid search parameters to optimize the performance of support vector machine, the support vector machine based on Particle Swarm Optimization can be used to predict the collision hazard identification of unmanned vessels ,and it has a good performance.

1. Introduction

The recognition of the safety situation of unmanned vessels is an important guarantee for the navigation of ships in the river. The collision risk of ships is one of the important indicators of the safety situation of unmanned vessels. According to the statistics of some Chinese provinces and local maritime bureaus A total of 165 water traffic accident data released on the official website, the collision accident accounted for as high as 68%, ranked first in water traffic accidents. Based on this data, it is particularly important to study the collision risk of unmanned ships.

Xue [1] adopted DCPA and TCPA weighted method to calculate the ship collision risk when multiple ships would encounter, and then determined the best path for collision avoidance decision. In 2000, Fan [2] put forward an evaluation method called Hazard Index Method, using the gas equation to get the collision accident indexes in all regions of the route, combined with the rest of the risk indicators affecting the normal operation of the ship and the ship flux weighting to get the ship Real-time risk index on a particular route. Yang [3] proposed a fuzzy Bayesian approach to the monitoring of shipping safety status. Taking the state as a node and the Bayesian network reasoning, the integrated safety probabilities and the risk factors of the ship under different environmental nodes were obtained Probability value. Ao [4] put forward the establishment of navigation safety assessment index system, the use of fuzzy comprehensive evaluation method to achieve the definition of different levels of risk factors.

Based on the results of previous studies, this paper proposes to evaluate the collision risk of unmanned ships by using support vector machines optimized by particle swarm optimization. The basic concepts and calculation models of collision risk are studied. The main factors that affect the risk of collision are analyzed, and the risk of collision under different conditions of the ship is obtained by fuzzy comprehensive evaluation. Based on this, particle swarm optimization is used to model support vector machines Parameter optimization, and the data regression results and support vector machines, grid-based support vector machines and genetic algorithms based SVM classification regression results were compared.

2. Sample data extraction and risk assessment

2.1. Sample data extraction

The sample data used in this paper is from 112 Chinese-made survey reports of collision between two ships issued by various provinces and local maritime authorities in China during 2004-2015. The report details the size information, speed of navigation and navigation position of two ships colliding with each other. Weather and water conditions information. Static information such as the size of inland river unmanned ships and ship encounters can be obtained by AIS; the dynamic speed of ship and ship encountering ship and heading and other dynamic information are obtained through information fusion of AIS and radar ; Real-time weather and water status information forms the environmental information on ship encounters through the Electronic Chart System (ECDIS). The Japanese scholar Iwasaki Kawai [5] proposed that the influence of collision probability of DCPA (shortest encounter distance) and collision risk of TCPA (shortest encounter time) with fuzzy reasoning method is proposed, which has been generally recognized by the sailing session.

2.2. Fuzzy comprehensive evaluation method

In this thesis, fuzzy comprehensive evaluation method is used to evaluate the collision risk level of unmanned ships. The original index data that affects the collision risk of ships mentioned above are used and the membership functions of different factors are given.

2.2.1. DCPA risk membership function. According to the definition of DCPA, the greater the value of the DCPA, the lower the probability of collision between the two ships, and the less dangerous it is. DCPA's risk membership function[6], such as equation (1) :

$$r_{DCPA}=\begin{cases}1 & DCPA\le d_1\\ \frac{1}{2}-\frac{1}{2}[\frac{\pi}{d_2-d_1}(DCPA-\frac{d_2+d_1}{2})] & d_1<DCPA<d_2\\ 0 & d_2\le DCPA\end{cases} \quad (1)$$

In this case, d_1 is the safe meeting distance of the ship, d_2 is the absolute safe distance of the ship, and d_1=d_2 in general.

2.2.2. TCPA risk membership function. TCPA can be expressed as the urgency of unmanned ships meeting the danger of collision with ships. In the event of a minimum encounter distance, TCPA indicates the time required for the collision between the two if the unmanned vessel took no collision avoidance measures. It can be seen that the smaller TCPA, the greater the risk of collision. TCPA membership function of the degree of risk [7] as formula (2) :

IOP Conf. Series: Journal of Physics: Conf. Series **1061** (2018) 012001 doi:10.1088/1742-6596/1061/1/012001

$$r_{TCPA}=\begin{cases}1 & t_{TCPA}\le t_1\\ [\frac{t_2-t_{TCPA}}{t_2-t_1}]^2 & t_1<t_{TCPA}<t_2\\ 0 & t_{TCPA}>t_2\end{cases} \tag{2}$$

$$t_1=\begin{cases}\frac{\sqrt{d_1^{\,2}-DCPA^2}}{V_R} & DCPA\le d_1\\ \frac{d_1-DCPA}{V_R} & DCPA>d_1\end{cases},\quad t_2=\frac{\sqrt{d_2^{\,2}-DCPA^2}}{V_R} \tag{3}$$

V_R is the relative speed of the target ship.

2.2.3. Collision risk assessment. Based on the fuzzy comprehensive evaluation method, the weighted expression of five factors that defines the collision risk of unmanned ships[8]:

$$CRI=w_{DCPA}r_{DCPA}+w_{TCPA}r_{TCPA}+w_{D_r}r_{D_r}+w_Q r_Q+w_K r_K \tag{4}$$

Among them, according to different factors, the degree of impact on the ship collision risk is different, based on the three methods of evaluation index weight: the Dofei method, expert survey method and AHP. This article defines the different influencing factors as :

$$W=\{w_{DCPA}=0.36, w_{TCPA}=0.32, w_{D_r}=0.14, w_Q=0.1, w_K=0.08\} \tag{5}$$

In this thesis, the collision risk is divided into three different stages, and the unmanned ships in different stages should be adjusted to varying degrees.

(1)Collision hazard level three

CRI is in the range of 0 to 0.4. The automatic control center of the unmanned ship is provided with ship data meeting with the ship and real-time tracking of the target ship, and real-time data are returned to re-evaluate the collision risk.

(2)Collision hazard level two

CRI is in the range of 0.4 to 0.7. Said to unmanned automatic control center for risk warning, apply for unmanned craft autopilot heading fine-tuning, return data for real-time risk assessment.

(3) Collision hazard level one

CRI is in the range of 0.7 to 1. Said the automatic control center for unmanned vessels to apply for emergency braking, this route is defined as not feasible, to re-routing to avoid collision.

3. Support vector machine

SVM (Support Vector Machine) model has great advantages in pattern recognition such as small sample, high dimension data and non-linearity. It is a modeling model based on VC dimension theory and minimization of structural risk, and can be solved under certain circumstances. Dimension disaster, over learning, owe learning and other machine learning some shortcomings. In practice, the choice of SVM parameters directly determines the performance of the whole model, and the choice of parameters is often based on actual experience, that is, through a large number of experiments choose the optimal parameters.

Grid search parameter optimization is known as a primitive programming approach to solve constrained nonlinear extremum problems [9], similar to the "exhaustive method." The optimization method is to divide the sample into grids within a certain range and then traverse all grid points to find the optimal parameters. However, due to the low classification accuracy of each parameter group in the grid, it is a waste of time to traverse the entire region when the accuracy of classification in a particular parameter group is high.

GA (Genetic Algorithm) has strong robustness and global optimization search ability [10]. It is a randomized search algorithm based on Darwin's theory of evolution and Mendel's genetic variation theory. Including coding, population initialization, selection, cross, mutation and other steps. The algorithm needs to acquire parameters such as the crossover rate and mutation rate, and the choice of these parameters seriously affects the quality of the solution. At present, the choice of these parameters is mostly based on experience. There is no feedback information that can make timely use of the network.

Particle Swarm Optimization (PSO) is a kind of optimization algorithm for simulating swarm intelligence . It uses some particles without mass and volume as individuals to initialize the particles randomly by optimizing the search to form a population, so that the particles The group presents complex features, changing the position of each particle, the entire algorithm iteratively. Particle swarm optimization, as an efficient optimization tool, not only has excellent global search ability and fast convergence speed, but also has a good effect on identifying collision risk of inland river unmanned ships. The parameters can get accurate danger through training Degree level.

We define the input sample dataset as T, and the steps of hazard algorithm based on PSO-SVM are as follows:

① Divide sample data set T into training data set T1 and test data set T2, and separate the first rank of danger levels of data sets T1 and T2 as the identification sample set, leaving the remaining 8 data sample elements, Data normalization;

② A set of SVM classifier parameters can be obtained by using each support vector in the normalized training dataset T1 (including the representative samples and the element samples) to form a particle and form a particle population X;

③ The population of particles set initialization parameters C1, C2, the initial velocity matrix, the optimal individual particle initial individual position and the global optimum position;

④ (6) is the fitness function of particle swarm, calculate the fitness function value of each particle in particle swarm;

$$F(x) = \frac{1}{m}\sum_{i=1}^{m}(f_i - y_i)^2 \tag{6}$$

⑤ According to the value of the fitness function, the particle's best position and the global optimum position are adjusted, and the state of the example is updated to obtain a new set of SVM classifier parameters;

⑥ When the fitness function value meets the requirement or reaches the maximum number of iterations, proceed to step ⑦, otherwise, return to step ④;

Using the best parameters for SVM network training, network prediction;

Table 1 is the simulation of the input data sample profile.

Table 1. Sample input data

Risk rating	CRI	DCPA	TCPA	Relative distance	Target ship Angle	Vt/Vo	Vr	Cr
1	0.93516	1.8	3.24	2.1	60	0.98019	20	0.99
2	0.43792	0.84	12.0	1.77	35	0.69565	7.73	6.56
3	0.30525	1.01	11.4	1.69	40	0.47916	7.08	3.24

4. Comparison and analysis of Simulation results

Taking the continuous CRI value of Table 1 as output data, 400 data samples of 500 sample data are taken as training samples and 100 data yang samples as test sample. Using the discrete Risk rating in the sample data as output data, the input data is the remaining attribute data values in the sample data. Taking 250 data samples from 500 sample data as training samples and 250 data yang samples as test samples. Comparisons of classification regression results of four different SVMs.

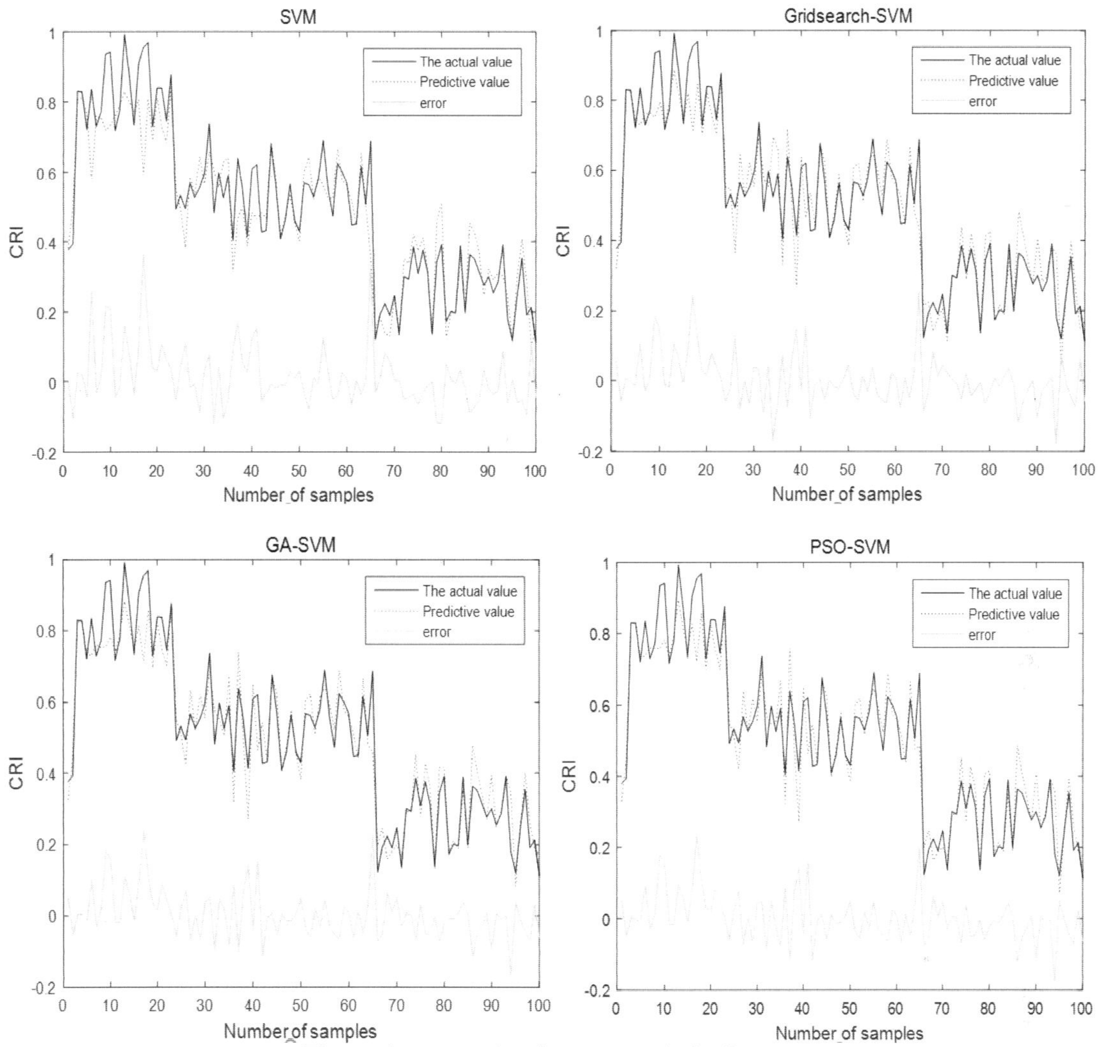

Figure 1. Regression forecast analysis diagram

Table 2. Regression data analysis

	Best c	Best g	MSE	Mean squared error	Regression
SVM	2	1	0.1	0.0419285	0.850686
GridSearch-SVM	8	1.442	0.04	0.0281851	0.895708
GA-SVM	9.4316	1.6682	0.03	0.0255490	0.906331
PSO-SVM	12.4562	1.7090	0.04	0.0237895	0.912771

Table 3. Multiple hazard level forecast results

	SVM	GridSearch-SVM	GA-SVM	PSO-SVM
The first time	%93.6	%96.4	%96	%97.6
The second time	%93.6	%96.4	%97.2	%98
The third time	%93.6	%96.4	%96.8	%98.4
Average value	%93.6	%96.4	%96.7	%98

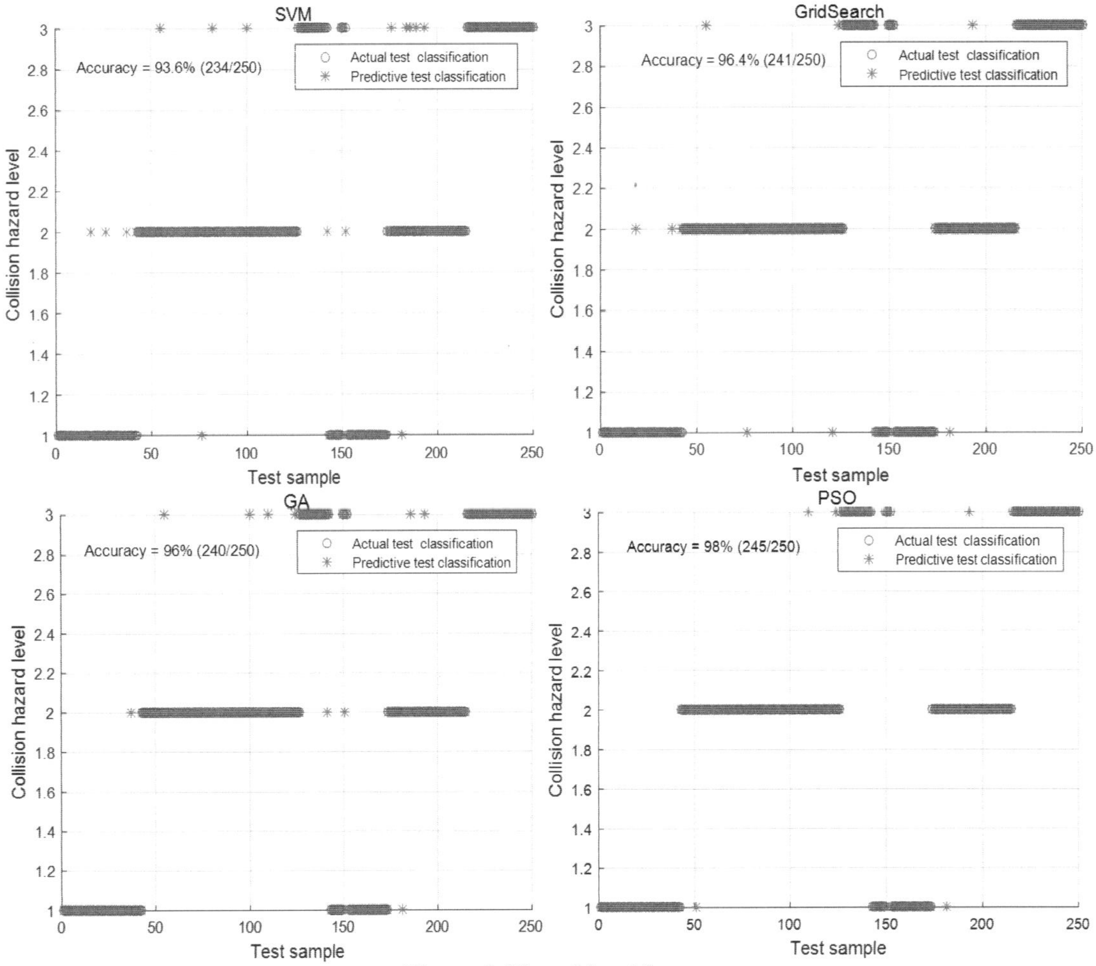

Figure 2. Hazard level forecast

Regression forecast analysis diagram is shown in Figure. 1 and Hazard level forecast is shown in Figure. 2. Table 2 shows the support vector box parameters optimized under different parameters. According to the above simulation results, the PSO-SVM has some advantages over the other two algorithms in predicting the accuracy of CRI regression, which greatly improves the classification accuracy of SVM for unmanned vessel's input data. Moreover, the PSO-SVM has the highest average among the results of the three-category classification, and the results of simulation classification in each group are superior to those of the other three SVMs. The results are shown in Table 3. Therefore, the PSO-SVM has a good recognition effect on the hazard degree classification of inland river unmanned vessels.

5. Conclusion

This article explores the classification of unmanned vessels in river collision hazard classification and forecasting problems. The system of unmanned ships is used to acquire the relevant data that affects the safe driving of the ship. The classification of ship collision hazard classification is carried out by using SVM based on Particle Swarm Optimization. After a large number of simulation experiments and comparison of simulation results, we can clearly see the advantages of PSO in global search ability, and can build SVMs with better classification and recognition effects, and improve the classification accuracy of support vector machines.

Acknowledgments
This work was supported by National Science Foundation of China (NSFC) under Grand 51479159 and the Fundamental Research Funds for the Central Universities under 185209005.

References

[1] Cai.Y.Z, Wei.Y, Sun.M. "Multi - ship Collision Avoidance Simulation Based on Collision Avoidance Ship Key Algorithm" . Journal of Dalian Maritime University:2014, 40(1):13-16.
[2] Fan. Y.T. "Traffic Safety Evaluation Method for Inland Waterways - Dangerous Index Method ".Wuhan Jiaotong University:2000.3.
[3] Yang.J, Zhang.Q.N. "Research on Shipping Safety Status Monitoring Based on Fuzzy Bayesian ".Wuhan University of Technology:.2016.04.018.
[4] Ao.J, Hu.Q.Y, Zhao.R.Y. "Analysis of real time safety index in navigable waters ".Shanghai maritime university. Ship sea engineering :2008.8.
[5] A.N.Cockeroft. "The circumstance of sea collision". The journal of Navigation: 1989:212-221.
[6] Li B, Pang F W. "An approach of vessel collision risk assessment based on the D–S evidence theory". Ocean Engineering: 2013, 74(7):16-21.
[7] Ren Y L, Mou J M, Yan Q X, et al. Study on Assessing Dynamic Risk of Ship Collision[C]// International Conference on Transportation Information and Safety. 2011:2751-2757.
[8] Xin.H, Zhou.Z.S, lv.A.Q,. "Shipwreck Risk Early Warning Based on Fuzzy Comprehensive Evaluation Method ". Chinese Journal of Safety Science: 2011, 21(4):115-120.
[9] Li.Q.Y ,Zhou.H, Lin.A.P. "Gray Melting Point Prediction Based on Grid Search and Support Vector Machine".Journal of Zhejiang University:2011:45-12.
[10] Chen.W.G, Ten.L, Liu.J, Peng.S.Y, Sun.C.X. "Prediction Model of Transformer Winding Hot Spot Temperature Based on Genetic Optimization Support Vector Machines".Transactions of China Electrotechnical Society:2014:29-1.

Starting Point Independent Iterative Learning Algorithm on LTI System with Non-zero Initial Errors

Byungyong You[1] and Jinwook Kim[2]

[1]School of Mechanical and Automotive Engineering, Kyungil University
50 Gamasil-gil, Hayang-eup, Gyeonsansi, Gyeongbuk, 38428, Republic of Korea
[2]Korea Electrotechnology Research Institute12, Bulmosan-ro 10 beon-gil, Seongsan-gu, Changwon-si, Geongsangnam-do, 51543, Republic of Korea

E-mail: zealot@kiu.kr

Abstract. In this paper, we propose a starting point independent iterative learning algorithm. This algorithm does not need to train iterative learning control (ILC) algorithm each time when initial condition changes. If the plant is sufficiently stable, the output trajectory converges to the desired trajectory exponentially fast. Regardless of learning rule type, this algorithm can be used and has a same convergence rate. So in case of the multiple starting and goal points application problems, it can be used with significant reduced computational cost. In fact, if there are M starting points and N goal points, the conventional ILC algorithm needs MN training for generating whole trajectories, however proposed algorithm needs only N training. In order to expand our algorithm to the multi-dimensional point-to-point (PTP) problem, we propose a learning rule for PTP problem, then shows its convergence.

1. Introduction

The ILC is used widely in real world applications that perform the same tasks repetitively. The basic idea of ILC is that the information obtained from previous trial is utilized to improve the control input for next trial. ILC has many advantages because it requires no or less plant information and it is robust for repetitive disturbances and uncertainties in a system. However, ILC generally requires the initial condition at each iteration and it should be the same and equal to that of the desired trajectory[1]-[6]. This requirement significantly limits ILC applications in the real field. Therefore there have been substantial research efforts in ILC with non-zero initial errors.

Heinzinger et al. [7] has shown that the output errors are bounded in the case of bounded initial errors. But there is no information on parameters on how to adjust the size of the bounds. Lee et al. [8] proposed a PD-type ILC algorithm on LTI system with parameter which reduces the non-zero initial errors and exponentially converges to the desired trajectory. However, it needs to train ILC every time whenever the initial condition changes. Therefore, in this paper, we propose a starting point independent iterative learning algorithm (SPIILA) on LTI system with non-zero initial errors. This algorithm does not need to train ILC algorithm each time when initial condition changes. If the plant is sufficiently stable, the output trajectory converges to the desired trajectory exponentially fast. So in the case of the multiple starting and goal points application problems, it can be used with significantly reduced computational cost. In fact, if there are M starting points and N goal points, the conventional ILC algorithm needs MN training for generating whole desired input trajectories but proposed algorithm needs only N training.In order to expand our algorithm to the multi- dimensional Point-to-Point (PTP) problem, we propose a learning rule for PTP, then shows its convergence.

This paper is organized as follows. Section 2 describes the ILC with non-zero initial errors. Section 3 details the usage and properties of the main theorem and propose the learning rule for PTP control problem and show the proof of its convergence. Section 4 shows computer simulation results. Finally, the conclusion and future works are given in Section 5.

In the following, $||x||$ denotes the Euclidean norm of a vector x. For a matrix A, A^T denotes its transpose and $||A||$ denotes its induced matrix norm. When A is a square matrix, $\lambda(A)$ is an eigenvalue of A. I_n denotes the $n \times n$ identity matrix. As in the notation $u_k(t)$, the suffix k is employed to denote the iteration number.

2. ILC with non-zero initial errors

Consider the LTI system described by

$$\dot{x}(t) = Ax(t) + Bu(t)$$
$$y(t) = Cx(t) \tag{1}$$

where $x \in R^n$, $u \in R^r$ and $y \in R^r$ denote the state, input and output respectively. A, B and C are matrices with appropriate dimensions and it is assumed that CB is non-singular. Then ILC algorithm for the system (1) can be described as follows [9]

$$u_{k+1}(t) = u_k(t) + \gamma\delta\dot{y}_k(t) \tag{2}$$

where

$$\dot{x}_k(t) = Ax_k(t) + Bu_k(t)$$
$$y_k(t) = Cx_k(t)$$
$$y_d(t) = Cx_d(t) \tag{3}$$

and

$$\delta y_k(t) = y_d(t) - y_k(t). \tag{4}$$

$$\text{If } ||I_r - \gamma CB|| \le \rho < 1 \tag{5}$$

and

$$y_k(0) = y_d(0) \equiv 0, k = 0,1,2,\cdots. \tag{6}$$

It is already known that by using the learning rule (2), the error between $y_k(t)$ and $y_d(t)$ approaches zero as $k \to \infty$. In the following, we want to discuss the case when (6) does not necessarily hold. The following Theorem 1, 2 describe the output trajectory when we use D-type and PD-type learning rule for the system (1) with same non-zero initial errors at each iteration. The proof of Theorem 1, 2 were provided by Lee et al. [8].

Theorem 1: Suppose that γ is chosen such that (5) holds, and that the D-type learning rule (2) is repetitively applied to (1). If the initial errors remain the same at each iteration, i.e. $x_k(0) = x_0, k = 0,1,2,\cdots$, then

$$\lim_{k\to\infty} y_k(t) = y_d(t) + Cx_0 \tag{7}$$

Theorem 1 shows that if the initial errors at each iteration remain the same, the output trajectory follows the desired trajectory with the offset of the initial errors. Now consider the following learning rule

$$u_{k+1}(t) = u_k(t) + \gamma(\delta\dot{y}_k(t) - R\delta y_k(t)). \tag{8}$$

Theorem 2: Suppose that (5) holds and the PD-type learning rule (8) is applied to (1). If the initial errors remain the same at each iteration, i.e. $x_k(0) = x_0, k = 0,1,2,\cdots$, then

$$\lim_{k\to\infty} y_k(t) = y_d(t) + e^{Rt}Cx_0 \tag{9}$$

Theorem 2 implies that if R is chosen such that $\lambda(R) < 0$, the learned control input enables the system to possess an asymptotic tracking capability even in the face of non-zero initial errors.

IOP Conf. Series: Journal of Physics: Conf. Series **1061** (2018) 012002 doi:10.1088/1742-6596/1061/1/012002

3. PTP problem and convergence when changing starting point

The conventional ILC focuses on tracking the desired trajectory for the fixed starting and goal points. However, it needs to train ILC when the initial error exists or user changes the initial condition arbitrarily. In this section we are going to investigate the usage of the desired input trajectory with the changing initial points. If this algorithm works, the pre-calculated desired input trajectory can be used to the changing points situation widely.

Theorem 3: Suppose that γ is chosen such that (2) holds, and suppose that the desired input trajectory when $x_d(0) = x_{d0}$ is applied to (1). If $x_k(0) = x_0, k = 0,1,2,\cdots$, and system matrix A is sufficiently stable then the output trajectory error decreases exponentially fast.

Proof:

We can describe the $y_k(t)$ from (1)

$$y_k(t) = Ce^{At}x_0 + C\int_0^t e^{A(t-\tau)}Bu_k(\tau)d\tau. \tag{10}$$

By assumption

$$y_d(t) = Ce^{At}x_{d0} + C\int_0^t e^{A(t-\tau)}Bu_d(\tau)d\tau. \tag{11}$$

If we substitute $u_d(\tau)$ for $u_k(\tau)$ into (10), then

$$\begin{aligned} y_k(t) &= Ce^{At}x_0 + C\int_0^t e^{A(t-\tau)}Bu_d(\tau)d\tau \\ &= Ce^{At}x_0 + y_d(t) + Ce^{At}x_{d0}. \end{aligned} \tag{12}$$

As a result, we can obtain the following error dynamics.

$$e_k(t) = y_k(t) - y_d(t) = Ce^{At}(x_0 - x_{d0}). \tag{13}$$

According to the assumption, if matrix A is sufficiently stable, the output trajectory error decreases exponentially fast. ■

Theorem 3 shows that we can change the starting point easily with non-zero initial errors. Also, this algorithm works for any type of learning rule that is applied to the system (1). According to (13), the convergence rate depends on the eigenvalues of the system matrix A. If $\lambda(A) \ll 0$, the output error becomes small. If the final time T becomes large, the output error also becomes small. Although the initial errors vary, this algorithm shows a same convergence rate to the various initial errors. However, in this algorithm, the system matrix A has to be a Hurwitz for convergence and there is no parameters to adjust for convergence rate. If we want to adjust the convergence rate, we have to move the pole location of the system using by feedback information. But, in order to use the feedback information, we have to know the precise system information. Therefore it is recommended to use this algorithm for a stable plant.

Let's find out the total computation cost. If there are M starting points and N goal points, conventional ILC needs MN desired trajectories to cover whole trajectories. However, if we use the proposed algorithm SPILLA, we can cover the whole trajectories with only N desired trajectories. That is, the previously calculated values can be reused. Therefore the total computation cost can be reduced efficiently. In case of $M \gg N$, this algorithm will be very effective.

Following example describes the SPILLA

Example 1. Consider the system (1) with

$$A = \begin{bmatrix} 0 & 1 \\ -1 & -2 \end{bmatrix}, B = \begin{bmatrix} 0 \\ 1 \end{bmatrix}, C = [0 \quad 1].$$

Suppose the desired trajectory is given as

$$y_d(t) = 12t(1-t), 0 \le t \le 1. \tag{14}$$

The P-type learning rule $u_{k+1}(t) = u_k(t) + \gamma\delta y_k(t)$ is used, value of γ is 0.7 and the iteration number is 50. Figure 1 shows the output trajectory with non-zero initial error converges to the desired trajectory when we use the SPILLA. The dotted line is a trajectory of conventional ILC when the initial error is zero and it means that 50 iterative training have been done. On the other hand, the dashed line (Proposed) is trajectory of SPILLA when the initial error is 1, i.e. $y_k(0) = 1, k = 0,1,2,\cdots$, and it means it is possible to reuse the desired input trajectory of ILC when initial condition changes 0 to 1. Therefore, if we want to change the starting point, we can adjust it by using initial errors.

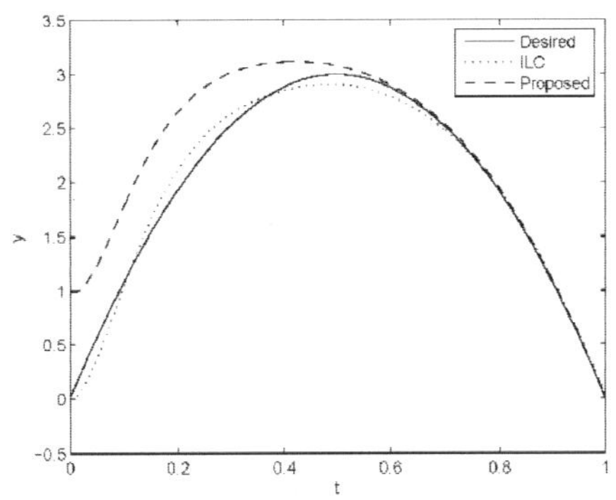

Figure 1. Example of SPILLA.

Now, we consider the PTP control problems. In the PTP problem which is no need to consider the whole trajectory, we can suggest the following learning rule.

$$u_{k+1}(0) = u_k(0) + \gamma\delta y_k(T), \tag{15}$$

where T is a final time of the PTP system. The convergence of the learning rule (15) is given in the following Theorem.

Theorem 4: Suppose that the learning rule (15) is applied to (1) and the initial errors are zero at each iteration, i.e. $x_k(0) = x_{d0}, k = 0,1,2,\cdots$,. If

$$\|I_r - \gamma Ce^{AT}B\|_\infty \le \rho < 1 \tag{16}$$

then,

$$\lim_{k\to\infty} y_k(T) = y_d(T). \tag{17}$$

Proof: Let $u_d(t)$ be a control input such that

$$\begin{aligned} y_d(T) &= Ce^{AT}x_{d0} + C\int_0^T e^{A(T-\tau)}Bu_d(\tau)d\tau \\ &= Ce^{AT}x_{d0} + Ce^{AT}Bu_d(0) \end{aligned} \tag{18}$$

The integral term can be represented as (16) because $u_d(t)$ exists only when t is zero. The proof is completed if it is possible to show that $lim_{k\to\infty} u_k(0) = u_d(0)$. We define $\delta u_k(t) = u_d(t) - u_k(t)$, then the main idea is to prove that $lim_{k\to\infty}\|\delta u_k(0)\|_\infty = 0$. Then it follows from (15) and (18) that

$$\begin{aligned} \delta u_{k+1}(0) &= u_d(0) - u_k(0) - \gamma\delta y_k(T) \\ &= \delta u_k(0) - \gamma[Ce^{AT}x_{d0} + Ce^{AT}Bu_d(0) - Ce^{AT}x_{d0} + Ce^{AT}Bu_k(0)] \\ &= \delta u_k(0) - \gamma[Ce^{AT}B\delta u_k(0)] \\ &= (I_r - \gamma Ce^{AT}B)\delta u_k(0) \end{aligned} \tag{19}$$

Taking the norm $\|\cdot\|_\infty$ on both sides of (19), we have

$$\|\delta u_{k+1}(0)\|_\infty \le \|I_r - \gamma Ce^{AT}B\|_\infty \cdot \|\delta u_k(0)\|_\infty \tag{20}$$

Since it is assumed that $\|I_r - \gamma C e^{AT} B\|_\infty \le \rho < 1$. Thus

$$\lim_{k\to\infty} \|\delta u_k(0)\|_\infty = 0 \tag{21}$$

$$\therefore \lim_{k\to\infty} y_k(T) = y_d(T) \tag{22}$$

This completes the proof. ■

4. Simulation results

We verify the main result for a single starting and goal point problem with initial errors and the simulation was performed in case of multiple starting and goal points problem. Finally, in order to verify the proposed algorithm is valid not only one-dimensional problem but also multi-dimensional problem, 3-dimensional PTP example is performed.

Figure 2 and 3 show that main result working for P and D-type learning rules, respectively. Consider the system (1) with

$$A = \begin{bmatrix} 0 & 1 \\ -1 & -5 \end{bmatrix}, B = \begin{bmatrix} 0 \\ 1 \end{bmatrix}, C = [0 \quad 1]. \tag{23}$$

Suppose the desired trajectory is given as

$$y_d(t) = 12sin^3(4t), 0 \le t \le 1$$

and let there are non-zero initial errors from -3 to 3. The $\gamma = 0.7, R = -5$, iteration numbers are 100, 15 for the P-type, D-type learning rule respectively. The solid line is desired trajectory and the dotted line is trajectory with ILC when the initial error is zero. Dashed lines are trajectories with the proposed algorithm when the initial errors are exists. We reuse the desired input information getting from the dotted line for SPILLA. According to SPILLA, trajectories with non-zero initial errors converge to the desired trajectory with same convergence rate.

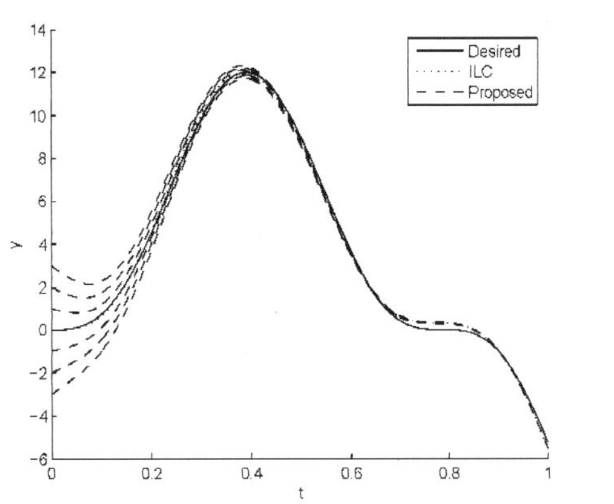

Figure 2. P-type learning rule for SPILLA

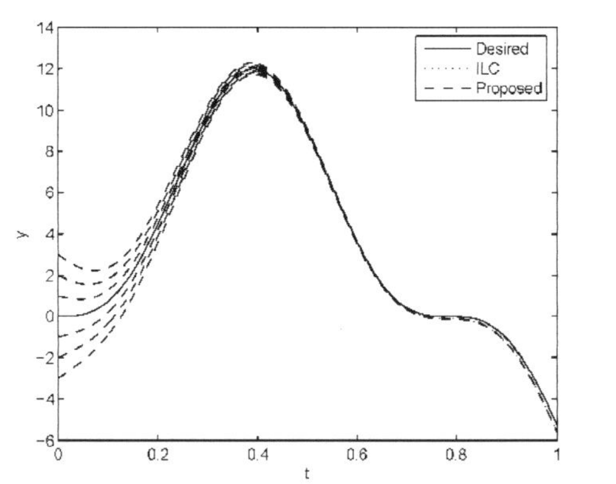

Figure 3. D-type learning rule for SPILLA

Figure 4. and 5. show the multiple starting and goal points problem. The system is same with above simulation. The desired trajectory is given as $y_d(t) = 12t(1-t), 0 \le t \le 1$. There are three starting points and goal points. As for whole 9 trajectories, 6 trajectories (dashed lines) are determined by SPILLA. It means only 3 trajectories are trained with the conventional ILC.

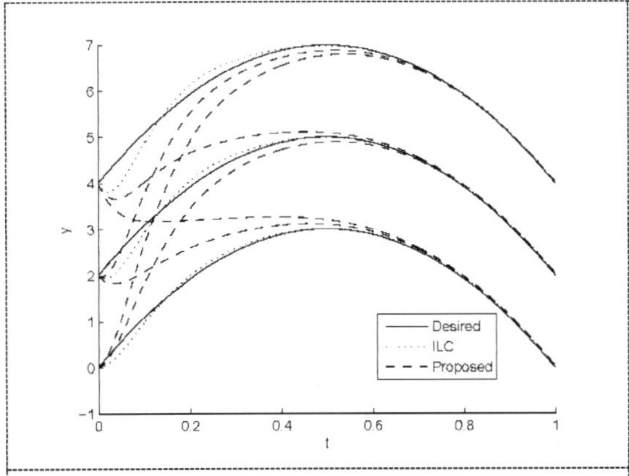

Figure 4. Multiple start/goal points (P-type)

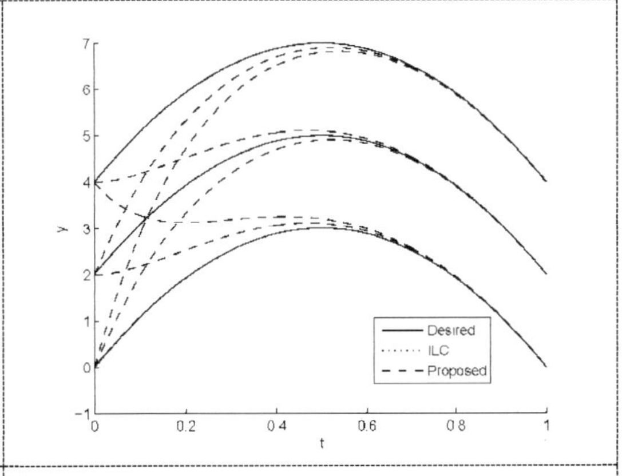

Figure 5. Multiple start/goal points (PD-type)

Now, we can expand this proposed algorithm to the multi-dimensional problem. Let's consider the 3-dimensional PTP problem.

$$\begin{bmatrix}\dot{x}_1\\ \dot{x}_2\\ \dot{x}_3\\ \dot{v}_1\\ \dot{v}_2\\ \dot{v}_3\end{bmatrix}=\begin{bmatrix}-1&0&0&1&0&0\\0&-1&0&0&1&0\\0&0&-1&0&0&1\\0&0&0&0&0&0\\0&0&0&0&0&0\\0&0&0&0&0&0\end{bmatrix}\begin{bmatrix}x_1\\x_2\\x_3\\v_1\\v_2\\v_3\end{bmatrix}+\begin{bmatrix}0&0&0\\0&0&0\\0&0&0\\1/m&0&0\\0&1/m&0\\0&0&1/m\end{bmatrix}\begin{bmatrix}F_x\\F_y\\F_z\end{bmatrix}+\begin{bmatrix}0\\0\\0\\0\\0\\-mg\end{bmatrix},\begin{bmatrix}y_1\\y_2\\y_3\\y_4\\y_5\\y_6\end{bmatrix}=\begin{bmatrix}1&0&0&0&0&0\\0&1&0&0&0&0\\0&0&1&0&0&0\\0&0&0&0&0&0\\0&0&0&0&0&0\\0&0&0&0&0&0\end{bmatrix}\begin{bmatrix}x_1\\x_2\\x_3\\v_1\\v_2\\v_3\end{bmatrix}, \quad (24)$$

where $m = 0.5kg, T = 5sec, g = \frac{9.8m}{s^2}, \gamma = 0.5$ for all dimensions and P-type learning rule is used. The starting point is (0,0,0) and goal point is (50, 50, 20). The initial force is (0,0,0) and using the learning rule given as (13), the initial force (F_x, F_y, F_z) is updated. The initial force effect to the system with very short time. Note that the given system (24) has 3-dimension and each dimension has a negative eigenvalue respectively. Therefore we can apply the SPILLA to this system.

Figure 6 shows the output trajectory when the PTP learning rule in (15) is applied in (24). The iteration number is 100. Then we can expect that the starting point can be changed with non-zero initial errors. Figure 7 presents that the each trajectory starting from several other points can converge to the same goal point by using SPILLA.

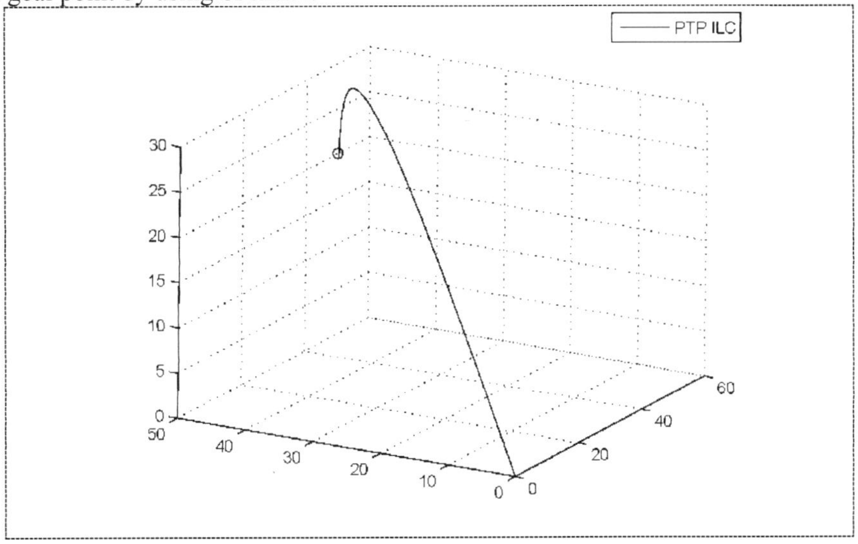

Figure 6. PTP problem with ILC

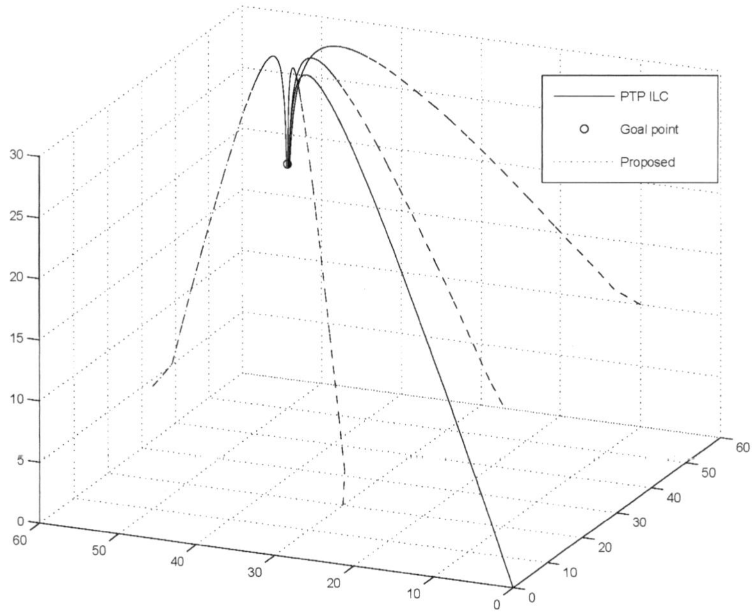

Figure 7. PTP problem with SPILLA

5. Conclusion

In this paper, we proposed a starting point independent iterative learning algorithm on LTI system with non-zero initial errors. This algorithm reduced the computation cost significantly. Also, the convergence rate for all starting points are the same. Then we suggested a PTP learning rule and showed its convergence. Validity of these algorithms are shown by computer simulation with multi-dimensional problem.

References

[1] Lee HS and Bien Z 2002 *Intelligent Auto. Soft Computing.* **8** 95-106
[2] Kuc TY and Lee JS 1992 *Automatica* **28** 1215-1221
[3] Chen Y, Wen C, Gong Z and Sun M 1999 *IEEE Trans. Auto. Cont.* **44** 371-376
[4] Park KH and Bien Z 2000 *Int. J. Cont.* **73** 871-881
[5] Sun M and Wang D 2003 *IEEE Trans. Auto. Cont.* **48** 144-148
[6] You B, Kim M, Lee D, Lee J and Lee J 2011 *Cont. Eng. Prac.* **19** 234-242
[7] Heinzinger G, Fenwick D, Paden B and Miyazaki F 1992 *IEEE Trans. Auto. Cont.* **37** 110-114
[8] Lee HS and Bien Z 1996 *Int. J. Cont.* **64** 345-359
[9] Arimoto S, Kawamura S and Miyazaki F 1984 *J. Robotic Sys.* **1** 123-140

IOP Conf. Series: Journal of Physics: Conf. Series **1061** (2018) 012003 doi:10.1088/1742-6596/1061/1/012003

Research Continued Mileage of Pure Electric Vehicles Based on Energy Management Control Strategy

Guo Minrui and Cheng Lei

College of mechanical and energy Engineering, Huanghuai University, Henan Zhumadian, China

814293109@qq.com

Abstract. In order to achieve the goal of increasing the mileage of pure electric vehicles, a pure electric vehicle was used as the research object, and the parameters of the battery pack were designed and matched, formulate the energy management control strategy based on threshold value, simulate the continuous driving distance of the vehicle with the help of ADVISOR simulation software, under the premise of not affecting the power of pure electric vehicles, the optimal threshold of energy management strategy that effectively increases the driving range of the vehicle is obtained, that is the battery pack's the state of charge is more than 45% and less than 50%, indicate that the energy management control strategy is correct and feasible.

1. Introduction

In 2015, China's production of new energy vehicles ranked first in the world, and foreign countries are also accelerating the development of electric vehicles. In December 2013, BYD officially announced that its electric bus K9 product was put into operation in London's public transport system in the United Kingdom. It can fly over 270 km without charge at one time. This is the first pure electric bus in London[1]. The traditional internal combustion engine has a mileage of more than 500km at one time. Some cars can even reach more than 1000km (such as Volkswagen Polo), Most electric cars have a range of between 100 and 300km in an ideal state, compared with the traditional car gap. While the famous American pure electric vehicle Tesla Model S is full of electricity, the mileage of one trip is 400~500km[2]. Compared with the driving range of pure electric vehicles both at home and abroad, compared with the traditional internal combustion engine vehicles, it can be seen that the driving range of pure electric vehicles still has more room for development. The pure electric car is powered by battery pack, the battery pack has limited energy. driving mileage is the key issue restricting the industrialization of pure electric vehicles. Therefore, it is necessary to start with the efficient use of vehicle energy, save and optimize the use of energy, and formulate a good energy management strategy to ensure that the driving motor can output effective torque and power rationally and prolong the driving range of pure electric vehicles. When the state of charge (SOC) of the battery pack decreases to a certain value, the motor will stop rotating due to the decrease of the battery terminal voltage, which will not only affect the service life of the motor and the battery pack, but also affect traffic safety[3]. How to develop a good energy management control strategy, choose to match the parameters of the battery pack to improve the driving range of pure electric vehicles, this is the key technology of pure electric vehicles.

2. Battery pack parameter selection

2.1. Battery type of choice

Lithium polymer battery energy density ratio, single high nominal voltage, safety performance, cycle life of more than 2000 times, and can discharge a larger current, high temperature performance, single cell capacity, no memory effect, light weight, pollution-free, environmentally friendly[4]. Such as BYD's electric bus K9, Tesla electric vehicles are used lithium-ion battery pack, indicating that it has the characteristics of other batteries do not have, so choose lithium polymer battery.

2.2. Battery energy and the number of choices

The total voltage of the battery pack should be slightly larger than the rated voltage of the drive motor to prevent the battery pack voltage from affecting the driving range of the electric vehicle. The rated voltage of the battery pack is 281.6V. The larger the vehicle driving range, the greater the battery power. Battery energy to ensure the driving range of the vehicle. The size of the driving range of electric vehicles is an important indicator of economic performance. Select the appropriate number of battery packs to ensure their driving range requirements. By the size of driving range of electric vehicles, can estimate the number of batteries required. In this paper, the specifications of the lithium polymer battery is 3.2V, 25Ah, according to the design requirements of electric vehicle in full charge of 40km / h speed uniform driving, the mileage is not less than 140km. The required power P is:

$$P = \frac{V_a}{3600\eta_t}\left(mgf + \frac{C_D A v_a^2}{22.15} \right) \tag{1}$$

In the above formula, m is the mass of the vehicle; g is the acceleration of gravity; f is the rolling resistance coefficient; C_D is the air resistance coefficient; A is the windward area; Va is the speed of the electric vehicle. Then the vehicle travels S = 140km at a constant speed with energy W:

$$W = Pt = P\frac{S}{Va} \tag{2}$$

Then the number of batteries required can be calculated by the following formula:

$$W = \frac{n\mu C E_0}{1000} \tag{3}$$

In the above formula, n is the number of needed batteries; μ is depth of battery discharge, the value is 0.8; E_0 is working voltage of the battery, the value is 3.2v; C is the capacity of the single battery, the value is 25Ah. After calculation, the number of single cells is 352, constitute a way for every 4 lithium polymer monomer cells in parallel to form a battery unit, and then in series with 88 battery cells, the total battery capacity of 100Ah. The battery pack parameters are shown in table 1.

Table 1. Battery pack parameters

parameters	the value
Single battery rated voltage	3.2（V）
Single battery rated capacity	25（Ah）
The number of single cells	352
Total battery capacity	100（Ah）

3. Energy management strategy formulation

Pure electric vehicles due to the limited number of battery pack, limited vehicle energy, the mileage will not reach the standard of traditional fuel vehicles[5]. There are two ways to solve this problem. Firstly, rely on the battery technology breakthrough to develop a larger total capacity of the battery pack, single battery terminal voltage is higher, the battery pack is better, faster charging battery, this is

IOP Conf. Series: Journal of Physics: Conf. Series **1061** (2018) 012003 doi:10.1088/1742-6596/1061/1/012003

more difficult to achieve. The second is to develop scientific and rational energy management strategy, reasonable allocation of battery energy, reduce energy consumption, improve energy efficiency. Improve driving range without affecting vehicle power and safety. At present, after optimization and redevelopment of the energy management control strategy, it can effectively increase the driving range of the vehicle.

The pure electric vehicle energy management system includes the estimation of SOC, energy management control strategy, regenerative braking energy recovery, etc. Energy management control strategy is a top priority. It not only controls the switching of various working modes of the vehicle, but also controls the rational distribution of energy flow. There are many types of energy management control strategies, such as logic threshold control strategy, fuzzy neural network control strategy, instantaneous optimization control strategy, global optimal control strategy, etc. Pure electric vehicles due to the use of a single energy source, energy management control strategy in two ways: First, the allocation of energy efficiency optimization vehicle, the second is the appropriate value of the SOC limit the motor torque output. Since pure electric vehicles only use batteries for power supply, the energy management control strategy is simple and the optimized follow-up mileage is obviously improved. General rules of the use of energy management strategies. According to the battery pack SOC value and drive motor efficiency, when the SOC value is different, the motor output power and torque changes. First set a good relationship between SOC and motor output power, when the SOC is reduced to a certain value, limit the motor power and torque output, to meet the dynamic and safety premise to improve the driving range. This strategy algorithm is effective and easy to operate, in practice, a large number of applications. The control method adopts the threshold control method., can effectively solve the motor and battery energy input and output problems. The method first sets a SOC threshold (between 30% and 40%) of the remaining energy. If the SOC estimation value is higher than the threshold, the output torque of the motor is not limited when the driver depresses the accelerator pedal. If the SOC estimate is below the threshold and even if the driver depresses the gas pedal, the energy management strategy limits the motor output torque, but in exceptional driving conditions such as overtaking or climbing, the motor output torque is relieved. Energy management control flow chart is as follows:

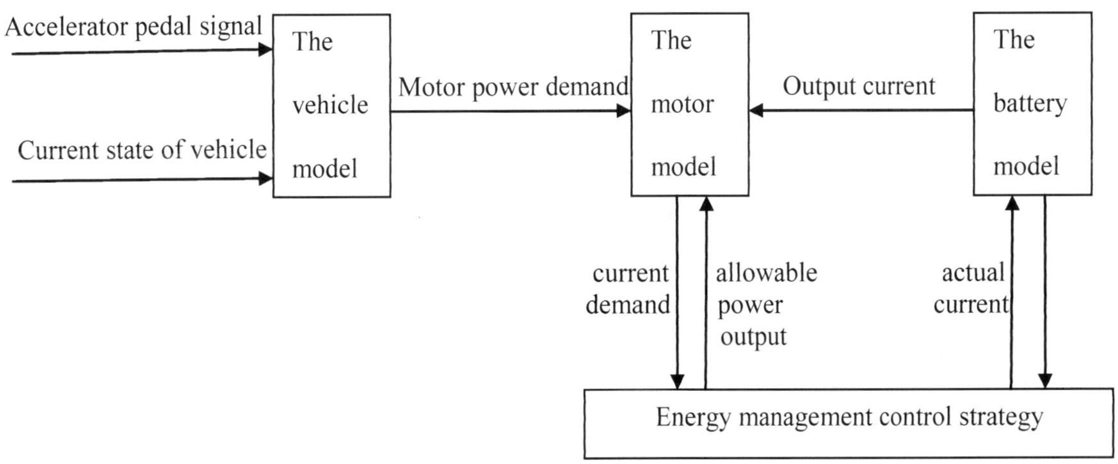

Figure 1. Flow chart of energy management control strategy.

In Figure 1, the acceleration pedal signal is converted into the demand torque value of the motor through the sensor, and the corresponding relationship between the acceleration pedal signal and the demand torque is made according to the driver's intention. Optimize the output torque of the motor through the energy management control strategy. According to the above control flow chart, the control strategy of the threshold value of the rules is established. When the pure electric vehicle is driving normally, the requirements of this strategy are as follows:

(1) if the battery group SOC is more than threshold value, the energy management strategy does not limit the output power and torque of the motor to ensure the power demand of the vehicle.

(2) if the battery group SOC is less than or equal to the threshold value, the energy management strategy limits the output power of the motor, which is 80% of the rated power.

(3) once in special conditions, such as driving state when overtaking or climbing hill, pure electric vehicles demand for motor torque is bigger, the energy management strategy relieves the motor torque limit, meets the needs of the vehicle with high torque, to ensure that the power performance and security.

4. Simulation analysis and experimental verification of vehicle continuous mileage

4.1. Input of battery pack and other parameters of the vehicle

The ADVISOR is a software for simulation analysis of traditional internal combustion engines, fuel cell vehicles, hybrid electric vehicles and pure electric vehicles. Travel distance with ADVISOR software simulation of vehicle, to reduce the electric car research and development cycle, compare the difference of driving range under different energy management strategies, first the input battery parameters and the other parameters, then select the simulation conditions.

4.2. Select simulation conditions

The ADVISTOR software offers three simulation scenarios: Standard Drive Cycle, Multiple Cycles and Test Procedures. Users can choose to simulate different operating conditions according to different requirements. With reference to the requirements of the Ministry of Industry and Information Technology of People's Republic of China , according to the actual conditions in the roads, choose the CYC-NEDC operating conditions under the standard cycle conditions to simulate the driving range. The relationship between speed and time of CYC-NEDC cycle is shown in Figure 2.

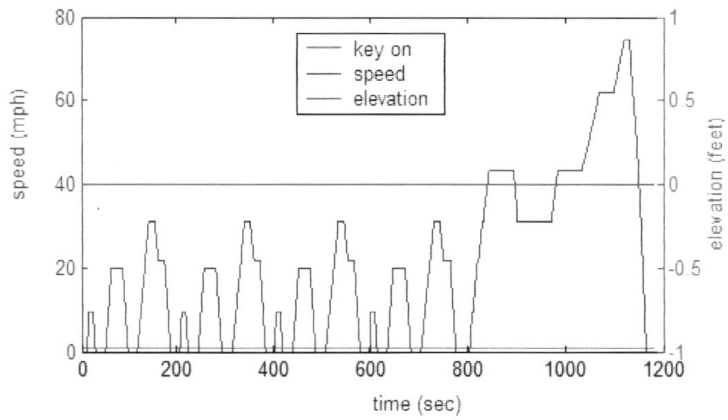

Figure 2. CYC-NEDC cycle conditions diagram

4.3. Driving range simulation results

ADVISOR software can be used to simulate the pure electric vehicle performance and mileage, etc, this part of the simulation includes two parts.

The first part need set the battery charged status (SOC) is 100%, the end of the discharging, SOC was 20%, running multiple NEDC driving cycles, the simulation results of the driving range of the pure electric vehicle, which determine the choice of battery parameters and the other parameters can meet the design index. The second part aims at increasing the mileage of continuous driving, and divides the total mileage into two stages. The initial stage is SOC from 100% to 50%, 45%, 40%,The second phase is SOC from 50%, 45%, 40% to 20%, energy management strategy limits motor output power, the sum of the two is the total driving range, compared with travel distance of the first part of each other, verify that the control strategy can increase the range of pure electric vehicles, learned that the energy control strategy is scientific and reasonable. Verify whether the control strategy increases the cruising range of pure electric vehicle and whether it is scientific and reasonable.

(1) Use the method of the first part, under the NEDC cycle conditions, the SOC value of the battery sets between 100% -20%, after simulation the driving range S is 147.8km, this driving range can reach the initial design specifications.

(2) Use the method of the second part, the threshold value SOC takes different values and limits the output torque of the motor when it is lower than the threshold value. Import the control policy into the ADVISOR's control policy library. In the first stage, the SOC value of the pure electric car battery pack is from 100% to 50%, and the mileage of this stage is 91.8km. The second stage sets the SOC value from 50% to 20%, and the mileage of this stage is 72.9km. The overall mileage of the vehicle S1 is 164.7km.

(3) Similarly, Use the method of the second part, In the first stage, the SOC value of the pure electric car battery pack is from 100% to 45%, and the mileage of this stage is 97.8km. The second stage sets the SOC value from 45% to 20%, and the mileage of this stage is 62.6km. The overall mileage of the vehicle S2 is 160.4km. S3 is 156.7km.Simulation comparison chart can be obtained by using different threshold values to simulate the mileage, as shown in figure 3.

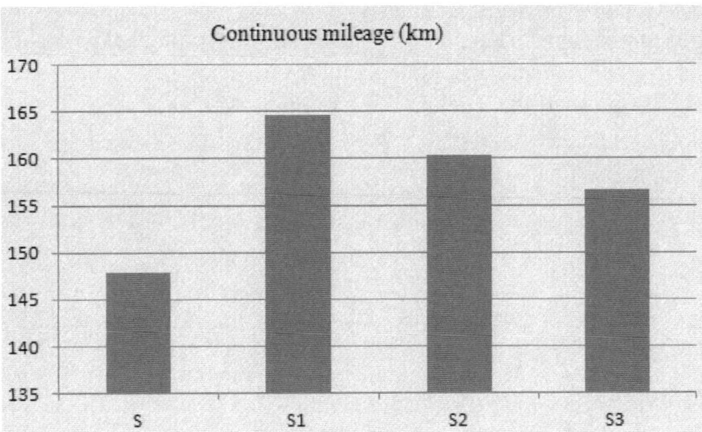

Figure 3. Compare the mileage simulation chart with different threshold values

According to figure 3, and collate the above data, under different SOC values, the following table 2 is obtained.

Table 2. Comparison of mileage simulation for different threshold values

	The SOC range		The driving distance
S	100%—20%		147.8
S1	100%—50%	50%—20%	164.7
S2	100%—45%	45%—20%	160.4
S3	100%—40%	40%—20%	156.7

After analyzing the above simulation results, S1, S2 and S3 adopted the energy management control strategy based on the threshold value, and the continuous mileage increased by 11.4%, 8.5% and 6.0%, respectively compared with those without the strategy. According to the above simulation results, in order to achieve the goal of increasing travel distance, at the same time balancing the power performance of vehicle, the torque of the motor limit point SOC values between 50% - 45%, that is, the threshold value of the energy management control strategy.

5. Conclusion

Different energy management control strategies will have certain impact on the driving range of pure electric vehicle. By adopting the NEDC cycle driving condition and setting the battery SOC range between 100% and 20%, the key point is to find the optimal threshold of different energy management control strategy, that is, the SOC value is between 50% and 45%, it can balance the power of the vehicle and the driving range, and compared with driving range of the new control strategy and the original control strategy, there has been some improvement in the mileage., indicate that the energy management control strategy is scientific and rational.

References

[1] Chen Xing, Xie Mingwei. Estimation of residual driving distance of electric vehicles with fuzzy energy consumption.[J]. *Journal of Henan University of Science and Technology*, 2017(2):28-34.

[2] Li Jinxia, Wu Qinmu. Estimation of SOC of lead-acid batteries for electric vehicles [J]. *Automation and instrumentation*,2015 (7) :221-222.

[3] Zhou Bin. Study on the calculation of SOC and continuous mileage of pure electric vehicle power batteries [D]. *Hefei University of Technology*, 2014.

[4] Du Shuang. Research on key technologies for energy management of dual energy source pure electric vehicles [D].*Jilin University*, 2015.

[5] Wang Guoqiang,Wang Zhixin. Research on optimization of new electric vehicle frame structure [J]. *Automation and instrumentation*,2016 (7) :67-69.

IOP Conf. Series: Journal of Physics: Conf. Series **1061** (2018) 012004 doi:10.1088/1742-6596/1061/1/012004

Calibration Method for Dynamic Deflection Measurement of Bridges by Photogrammetry

Huayang He, Yishu Zhou, Jinjin Cao and Jinning Zhang

Research Institute of Highway, Ministry of Transport, Beijing, China

ys.zhou@rioh.cn

Abstract. Heavy vehicular traffic and aggressive environmental conditions can cause unexpected bridge deterioration, thus, periodic inspections to identify and assess possible defects are required. One indicator is the amount of vertical deflection that occurs during loading. Some different types of transducers can be used to measure the deflection. One of them is the photogrammetry. It is a typical noncontact measurement method. Unfortunately, it is difficult to calibrate this type of instruments. The key point is the evaluation of the dynamic deflection measurement capability. A dynamic calibration system of bridge photogrammetric measurement system was composed of the motorized linear stage, LED target and computer. The Fréchet distance was used as a model to evaluate the dynamic deflection measurement capability of bridge deflection photogrammetry. The photogrammetric bridge deflection measuring instruments are products of BIOET, which are used to carry out the measurement experiment. The results show that this system can calibrate the dynamic deflection measurement error of photogrammetric bridge deflection measuring instruments. Fréchet distance can reflect the relationship between the curves. This evaluation method is more accurate than the current evaluation method based on the error of indication of maximum dynamic deflection.

1. Introduction

For the structural evaluation of bridges, different tests are usually performed to know the structural properties such as natural frequencies, dynamic responses, static deflections, dynamic deflections, stresses, and strains [1]. What is more, it is important to monitor the change of these properties as well as damages to evaluate the structural integrity of bridges. One of these properties is vertical deflection of bridge girders. Vertical deflection includes static deflection and dynamic deflection. Several different types of sensors can be used to measure the static deflection and dynamic deflection [2-3]. However, most of these sensors need to be installed under the bridges. Therefore, the tests would interrupt the traffic under the bridge for the setup of these sensors. Another approach to measure bridge static deflection and dynamic deflection is to use the noncontact measurement methods. Noncontact methods realize the measurement of the deflection of bridges without occupying space and interrupt the traffic under bridges. Thus, various methods such as Photogrammetry [1,4], Moiré [5-6], Laser Scanning [7-8], and Laser Doppler [9] have been proposed, utilized, and applied to measure real or model bridge static deflections and dynamic deflections. Close-range photogrammetry has been used to determine the three-dimensional geometry and deformation of bridges since the 1970s.

Close-range photogrammetry for bridge deformation measurement has many advantages. It can measure difficult-to-access bridges by taking pictures away from the bridge and record a large amount

IOP Conf. Series: Journal of Physics: Conf. Series **1061** (2018) 012004 doi:10.1088/1742-6596/1061/1/012004

of two-dimensional information in a short time. A basic bridge photogrammetric measurement system consists of a camera, accessories, LED targets, scale bars, and software for photogrammetric processing. However, calibrating this type of instruments is very difficult. The key point is the evaluation of the dynamic deflection measurement capability.

The camera photogrammetric calibration was in need of an automated system [10]. Calibrated devices consisting of the motorized linear stages can be used to calibrate static deflection, the error of indication is not more than 0.01mm [11]. However, it was difficult to calibrate the dynamic deflection measurement error. Dynamic Bayesian Network and Canonical Time Warping were used in motion retrieval to recognize the similarity of two motion curves [12]. The Fréchet distance was used by some researchers to measure the correlation between two polygonal curves [13]. And some used the Hausdorff distance metric to comparative analysis of dendritic architecture of identified neurons [14].In the signature verification method, the distance threshold was got by training the Siamese Network [15].

According to the needs of transportation industry, standard and verification regulation of the bridge photogrammetric measurement system are being made. A displacement device was designed. The displacement device was used to simulate the bridge vibration. The operation of the displacement device was recorded by a digital camera. Hereafter, the deflection distribution was evaluated by model, which was based on Fréchet distance. The results show that the model can be used to evaluate the results obtained using digital image technique.

2. Devices and methods

2.1. Displacement device

The displacement device simulating the bridge vibration process was composed of the motorized linear stage, LED target and computer (**Figure 1**). The motorized linear stage is a positioning device providing high precision linear motion with minimized runout. This behaviour is critical to the success of simulation. The LED target is fixed on the stage.

Figure 1. Displacement device

2.2. Dynamic deflection curves

As dynamic load test was performed to measure the dynamic deflection of the bridge, which was a prestressed continuous small box girder bridge with the span of 20 meters, the dynamic response of the structure under forced vibration was recorded. The load of moving vehicles would cause vibration, shock, and dynamic response.

A car drove through the bridge at a speed of 40km/h without any obstacles. Arranging the strain measuring points in the control sections to measure the dynamic strain of the bridge structure under dynamic loads. The testing data about the strain collected automatically by strain gauges, data acquisition instrument and computers. Dynamic strain on the experimental process, the dynamic vertical deflection signal was recorded, sorted, and data was analysed to determine the dynamic vertical deflection of the bridge structure under traffic loads. One of the measured dynamic deflection curves is shown in Figure 2.

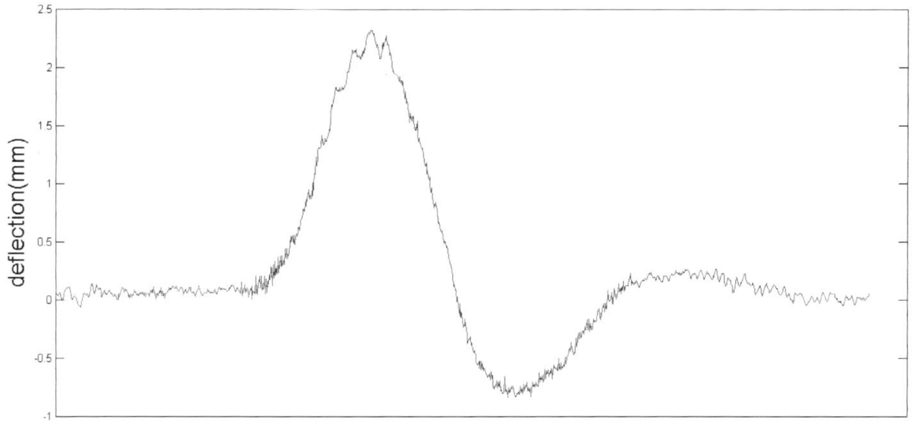

Figure 2. dynamic deflection curve

2.3. Dynamic Measurement Capability Assessment Model

The Fréchet distance is a popular and widespread distance measure for point sequences and for curves [16].Suppose two-tuple(S, d) is a metric space, and d is the metric function of S. The mapping $\gamma: [0,1] \rightarrow S$ in unit interval [0, 1] is continuous mapping, then γ is continuous curve in S. If $\tau: [0,1] \rightarrow [0,1]$ subject to three conditions: (1) τ is continuous. (2) τ is non-descending. (3) τ is surjection. This implies that τ is the reparameterized function of unit interval [0, 1], and $\tau(0) = 0$, $\tau(1) = 1$.

During the calibration of the photogrammetric bridge deflection measurement system. The dynamic deflection curve measured by the laser interferometer is called the standard curve, and the dynamic deflection curve measured by the photogrammetric bridge deflection measurement system is called the detected curve. Suppose the standard curve (sc) and the detected curve (dc) are the continuous curves in S. And $SC: [0,1] \rightarrow S$, $DC: [0,1] \rightarrow S$, then the Fr´echet distance between standard curve and detected curve defined as $F(sc,dc) = inf \max_{t\in[0,1]} \left\{d\left(sc(SC(t)), dc(DC(t))\right)\right\}$, d is the metric function of S. It is shown in Figure 3.

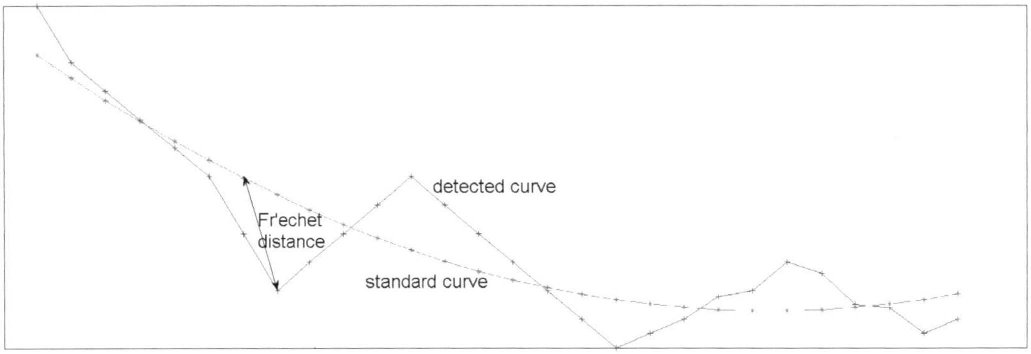

Figure 3. Fréchet distance

3. Results and discussion

The photogrammetric bridge deflection measuring instruments were products of BIOET, which were used to carry out the measurement experiment (As is shown in Figure 4).

Figure 4. measurement experiment

The error of indication of maximum dynamic deflection were calculated by $\frac{|SC_{max}-DC_{max}|}{SC_{max}} \times 100\%$, where SC_{max} is the maximum dynamic deflection of the standard curve, DC_{max} is the maximum dynamic deflection of the detected curve. In order to facilitate the analysis, the results of Fréchet distances was normalized.

The results presented in Figure 5 and Table 1 show that DC2 is more deviated from SC than DC1, the maximum dynamic deflection of DC4 appears at the end of the curve, and DC3 has the lowest similarity to SC. Fréchet distances accurately reflects the relationship between the curves. The Fréchet distance between DC1 and SC is smaller than the Fréchet distance between DC2 and SC. Fréchet distance between DC3 and SC is the maximum. However, the error of indication of DC2 is smaller than DC1. The error of indication between DC1 and SC is similar to the error of indication between DC4 and SC. This indicates that the error of indication cannot present the situation of the curves accurately.

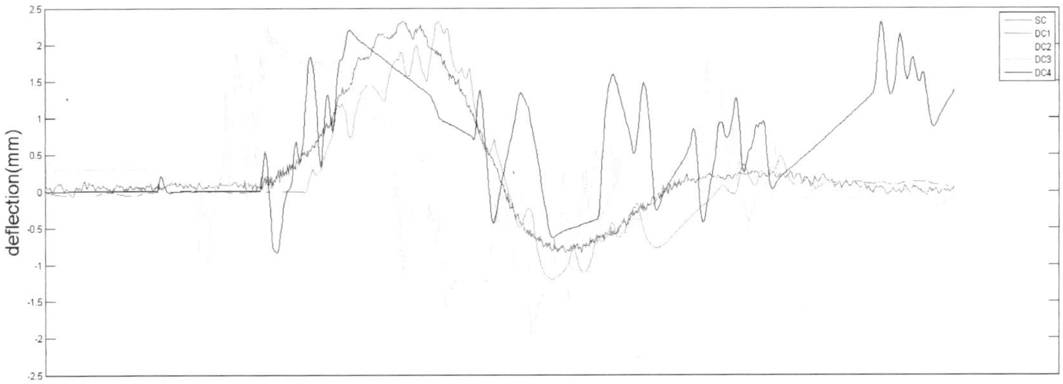

Figure 5. the experiment results

Table 1. the experiment results

Detected curve	Error of indication (maximum dynamic deflection)	Fréchet distance (normalization)
DC1	0.624%	1
DC2	0.163%	1.69
DC3	0.976%	3.48
DC4	0.608%	2.70

The Fr´echet distance is a curve comparison method rather than the error of indication, which indicates point-to-point comparison. As the Fr´echet distance was better suited for comparing results on their intrinsic structure, the Fr´echet distance and its variants can be used to compare the results obtained by photogrammetry technique and displacement device.

4. Conclusion

A dynamic calibration system for photogrammetric measurement of bridges was designed based on the motorized linear stages. And this system can calibrate the dynamic deflection measurement ability of photogrammetric bridge deflection measurement system.

Fréchet distance can reflect the relationship between the curves more accurately than the error of indication of maximum dynamic deflection. Compared to the error of indication, Fréchet distance is of more practical significance. All the work have built a solid base for more dynamic deflection test.

5. Acknowledgements

This work is supported by the Standardization Project of MOT (No. 2016-04-37) and Central Public-interest Scientific Institution Basal Research Fund (No.2017-9061). We thank Hongbo Guo for help with experiments.

References

[1] Yoneyama S, Kitagawa A, Iwata S, et al. BRIDGE DEFLECTION MEASUREMENT USING DIGITAL IMAGE CORRELATION[J]. Experimental Techniques, 2007, 31(1):34–40.

[2] Nassif H H, Gindy M, Davis J. Comparison of laser Doppler vibrometer with contact sensors for monitoring bridge deflection and vibration [J]. Ndt & E International, 2005, 38(3): 213-218.

[3] Guan S, Rice J, Li C, et al. Bridge deflection monitoring using small, low-cost radar sensors[C]//Structures Congress 2014. 2014: 2853-2862.

[4] Jiang R, Jauregui D V. Development of a digital close-range photogrammetric bridge deflection measurement system [J]. Measurement, 2010, 43(10): 1431-1438.

[5] Chen X, Chang C C. In-plane motion measurement by using digital sampling moiré method[C]//Sensors and Smart Structures Technologies for Civil, Mechanical, and Aerospace Systems 2016. International Society for Optics and Photonics, 2016, 9803: 980311.
[6] Tomita D, Murata Y, Fujigaki M. 3D Displacement Distribution Measurement Using Sampling Moire Method with Multiple Cameras[C]//Emerging Challenges for Experimental Mechanics in Energy and Environmental Applications, Proceedings of the 5th International Symposium on Experimental Mechanics and 9th Symposium on Optics in Industry (ISEM-SOI), 2015. Springer International Publishing, 2017: 187-194.
[7] Sithole G, Vosselman G. Bridge detection in airborne laser scanner data [J]. ISPRS Journal of Photogrammetry and Remote Sensing, 2006, 61(1): 33-46.
[8] Truong-Hong L, Laefer D F. Using Terrestrial Laser Scanning for Dynamic Bridge Deflection Measurement[C]//IABSE Istanbul Bridge Conference, Istanbul, Turkey, 11-13 August 2014. 2014.
[9] Miyashita T, Ishii H, Fujino Y, et al. Clarification of the effect of high-speed train induced vibration on a railway steel box girder bridge by monitoring using laser Doppler vibrometer [C]//Proceedings of Third International Conference on Bridge Maintenance, Safety and Management. 2015: 265-267.
[10] De Villiers J. Design and application of an automated system for camera photogrammetric calibration[D]. University of Cape Town, 2015.
[11] ZHOU Zhi-Chun. Research on Automatic Calibration System for Bridge Deflection Instrument [J]. Quality and Technical Supervision Research, 2015(1):32-33.
[12] Xiao Q, Siqi L. Motion retrieval based on dynamic bayesian network and canonical time warping[J]. Soft Computing, 2017, 21(1): 267-280.
[13] Alt H, Godau M. Computing the Fréchet distance between two polygonal curves[J]. International Journal of Computational Geometry & Applications, 1995, 5(01n02): 75-91..
[14] Mizrahi A, Benner E, Katz M J, et al. Comparative analysis of dendritic architecture of identified neurons using the Hausdorff distance metric.[J]. Journal of Comparative Neurology, 2015, 422(3):415-428.
[15] Bromley J, Guyon I, LeCun Y, et al. Signature verification using a" siamese" time delay neural network[C]//Advances in Neural Information Processing Systems. 1994: 737-744.
[16] Bringmann K, Mulzer W. Approximability of the discrete Fréchet distance[J]. Journal of Computational Geometry, 2015, 7(2): 46-76.

The Comparison between Dragonflies Algorithm and Fireflies Algorithm for Court Case Administration: A Mixed Integer Linear Programming

W Wongsinlatam[1] and S Buchitchon[2*]

[1]Faculty of Applied Science and Engineering, Khon Kaen University, Nong Khai Campus, Nong Khai, 43000, THAILAND, E-mail address : wullwon@kku.ac.th

[2]Faculty of Integrated Social Sciences, Khon Kaen University, Nong Khai Campus, Nong Khai, 43000, THAILAND

[*]Corresponding author E-mail address : suntareebu@kku.ac.th

Abstract. This article focuses on case assignment which is one of the most significant tools in case management to improve efficiency of judicial system. The effectiveness of judicial system depends greatly on the efficient and timely manner of court case operation, especially in criminal cases where suspects are hold in prison on remand. To ensure the rule of law, the preferred system in the court of justice is the one that is predictable and consistent. However, one of the building blocks to such goal is the nature of litigation which is unpredictably time consuming. In this paper, a Mixed Integer Linear Programming (MILP) is selected to address the issue of case assignment in the court of justice. The computational experiments are also reported to evaluate its solution. The two techniques of metaheuristic method are used to find optimal solution of the assignment problem that are called Dragonflies Algorithm (DA) and Fireflies Algorithm (FA). The result shows that DA is more effective in finding an optimal solution to the problem of court case assignment than of the FA.

1. Introduction

The preferred system in the court of justice is the one that is predictable and consistent to ensure the rule of law [1]. Even though justice administration and case assignment are routine works in judicial system, it is repeatedly reported that litigation is unpredictably time consuming . In the last decade, due to the change in constitutional law of Thailand which requires a full quorum of judges to be seated at a hearing, the caseload in the court of justice is dramatically increase. A proposal has been made to increase system efficiency through case management, court-annexed mediation, and digital audio testimony recording system [2]. However, the major challenges still remain which are to define the right case to be assigned to the right justice team and to allocate appropriate time for delivering the court decision. It is proposed in this article that a math model in the form of mixed integer linear programming MILP can be a helpful way to adopt a measure to develop the judicial system.

Generally, in planning the working system in an organization, the coordination of workers and tasks is a major consideration that reflects the efficiency of the company .The Mixed Integer programming (MILP) used in the examination proctor assignment problem by Takeshi Koide where staffs are assigned as proctors in the regular examination period [3]. Julia Rieck et al. [4] applied the

IOP Conf. Series: Journal of Physics: Conf. Series **1061** (2018) 012005 doi:10.1088/1742-6596/1061/1/012005

project scheduling problems subject to general temporal constraints using MILP model and domain-reducing processing techniques. David Bredström et al. [5] used MILP model for barge transport planning on the river Rhine .The planning solved the supply chain management of Omya's production of Norwegian high quality calcium carbonate slurry, supplied to European paper manufacturers . Christodoulos A .Floudas [6] applied the scheduling of chemical processing systems based on MILP approaches which improved the computational efficiency in the solution of MILP problems .However, the use of MILP model in juridical problem is still limited .

The main purpose of metaheuristic optimizations is to find optimal solution for NP-hard problem which the algorithm of metaheuristic optimizations is made of a random number that moves in a wide feasible solution to solve several optimization problems. Because some functions may have discontinuities, and thus derivative information is not easy to obtain, the metaheuristic optimization reduces the time in finding the exact solution or the good solution in the problem .Popular methods for assignment problem includes; Ant Colony Optimization (ACO) [7], Firefly Algorithm (FA) [8-9], and Dragonfly algorithm (DA) [10].

In this paper, the objective function to maximize the overall effectiveness is used in the MILP optimization model .The objective of assignment problem includes N cases that must be assigned to M teams where each team has the competence to do all cases .However, due to personal ability, case specification or other reasons, each team may spend different time for minimize or maximize with the objective of assignment problem .

The rest of the paper proceeds as follows :Section 2 presents the mathematical model for assignment problem of the case assignment in the court of justice .The objective function uses the MILP optimization by using two metaheuristic optimizations to find an optimal solution .The computational results and discussion about the effectiveness of the proposed metaheuristic algorithms are provided in Section 3 .Finally, section 4 presents the summarizes of our conclusions.

2. The Mathematical Model for Case Assignment Problem

2.1 The definition of the problem

In order to simplify the presentation of the approach, it is assumed in this the mathematical model that:

- There are fixed numbers of justice panels (3, 4, or 5) in one court and one judge can only be in one panel.
- All of the justice panels have the same expertise.
- One case is being assigned to only one panel.
- No interdependencies exist between each case.
- Each case required difference time limit depending on the nature of the case.
- The minimum effective time for delivering a decision for one case is one day. Whereas the maximum effective time for delivering a decision is 90 days which represents the longest time period where a person can be held in prison on remand.

The formulation of problem, C is index set for cases assignment, P is index set for justice panel, c is each case assignment, μ_c effective rate for case assignment, L_c and U_c are lower bounds and upper bounds on justice time. Next, MT is maximum time for justice deadline T_{cj} is assignment of case c to the right justice and X_{ij} is a binary decision variable. For determine the problem, a set $C = \{1, 2, .., n\}$ of n cases has been filed to the court of justice where a set $P = \{1, 2, .., m\}$ of m are available justice panels forming justice teams.

2.2 The objective function of problem

The MILP model is represented as follows:

$$\text{maximize} \qquad Z = \sum_{i=1}^{n}\sum_{j=1}^{m} \mu_i \cdot T_{ij} \tag{1}$$

Constraints:

$$\sum_{i=1}^{n} T_{ij} \le MT; \ \forall \ j \le m \tag{2}$$

$$T_{ij} - X_{ij} \cdot U_i \le 0; \ \forall \ i \le n, j \le m \tag{3}$$

$$T_{ij} - X_{ij} \cdot L_i \ge 0; \ \forall \ i \le n, j \le m \tag{4}$$

$$\sum_{j=1}^{m} X_{ij} \le 1; \ \forall \ i \le n \tag{5}$$

$$X_{ij} = \begin{cases} 1, \ if \ \ case \ i \ is \ selected \ \ to \ justice \ j; \\ 0, \ otherwise; \end{cases} \quad , \forall \ i \le n, j \le m \tag{6}$$

It is assumed that the effectiveness rates of cases administration are $\mu_1, \mu_2, .., \mu_n$, respectively. The objective function according to equation (1) for the above problem is defined to the overall effectiveness for the judicial mode. The constraint in equation (2) is the total time spent on a sequential of case for each justice team is less than or equal to MT. The constraint in equation (3) is called upper bound U_i for the time spent on each selected case i and it is no more than upper bound. The constraint in equation (4) is called lower bound L_i for the time spent on each selected case i and it is no more than lower bound. Define X_{ij} as a binary decision, let a number of 1 if case i is selected and assigned to justice team j and let a number of 0 if otherwise in constraint according equation (5) and (6).

2.3 The metaheuristic optimization

2.3.1 Inspiration. In applying the method, the cases are assigned to justice panels considering the maximum time available. The cases are sorted into a list with the amount of time allocated for each case. Then the first case in the list is assigned to the justice panel. The workstation used for carrying out the result is equipped with 2.40 GHz processor, 4.00 GB of RAM, and the system used Microsoft 7. Scilab free software was used to carry out simulation of the proposed algorithm. The number of cases for experimental parameters was 50 and the number of justice team for experimental parameters was 3, 4 and 5. The optimization method of maximizing effectiveness of case assignment in court of justice in this paper used two techniques of metaheuristic methods; Firefly algorithm (FA) and Dragonfly algorithm (DA). Adopting from the behaviour of fireflies, the flashing light is viewed as associated with the objective function to be optimized. Thus, makes it possible to formulate new optimization algorithm so called Firefly algorithm (FA). For the Dragonflies algorithm, the interesting fact about dragonflies is their unique swarming behaviour. Naturally, dragonflies swarm in hunting and migration. Such behavior can be depicted to form optimization algorithm called Dragonfly algorithm (DA).

2.3.2 The Firefly algorithm for problem (FA) and The Dragonfly algorithm for problem (DA). The FA technique was firstly introduced by Yang in 2008 [9]. The FA is generally used in solving continuous optimization in NP-hard problems. Algorithm 1 illustrates the FA pseudo code. The second metaheuristic method that is being used in this paper is DA. The DA is proposed firstly by Seyedali Mirjalili in 2016 [10] from the inspiration of static and dynamic swarming behaviors of dragonflies. These two swarming behaviours are very similar to the two main phases of exploration and exploitation in optimization using meta-heuristics. The objective swarming behaviors are reflected in the algorithm of DA. The algorithm of DA is presented in Algorithm 2.

Algorithm 1. Pseudo code of the FA

Define the objective function $f(x)$;
Initialize the firefly population $x = x_1, x_2, \ldots x_n$;
Define the light absorption coefficient γ;
for each firefly x_i in the population **do**
    Initialize light intensity I_i;
end
repeat
    for each firefly x_i in the swarm **do**
        for each other firefly x_j in the swarm **do**
            if $I_i > I_j$ **then**
                Move firefly x_i toward x_j;
            end
            Attractiveness varies with distance r via $\exp(-\gamma r)$;
            Evaluate new solutions light and update intensity;
        end
    end
    Rank the fireflies and find the current best;
until termination criterion reached;
Rank the fireflies and return the best one;

Algorithm 2. Pseudo code of DA

Initialize the dragonflies population;
Initialize step vectors
while the end condition is not satisfied
    Calculate the objective values of all dragonflies;
    Update separation weight, alignment weight, cohesion weight, food factor and enemy factor;
    Calculate separation, alignment, cohesion, food source, position of enemy of i-th individual;
    Update neighbouring radius;
    if a dragonfly has at least one neighbouring dragonfly
        Update velocity vector;
        Update position vector;
    else
        Update position vector;
    end if
    Check and correct the new positions based on the boundaries of variables;
end while

2.3.3 The metaheuristic optimization to solve problem. The proposed metaheuristic optimization to solve the problem is described as follows:

Input: Index set of C and P. For each case $c \in C$, effectiveness rate is μ_c and maximum time for justice team deadline is MT. Lower and upper bounds on decision times are L_c and U_c, respectively.

For assumption, the effectiveness rates for each case ($\mu_1, \mu_2, \ldots, \mu_n$) are in decreasing order.

Step 1: Set $c = 1$, and set $Z_c^* = T_{ci}^* = 0$, $MT_i = MT$, for each $c \in C$ and $i \in P$.

Step 2: Find justice $i \in P$, and MT_i if $U_c \leq MT_i$ then $T_{ci}^* = U_c$ and update $MT_i = MT_i - T_{ci}^*$. Next, go to step 4 and go to step 3 where otherwise.

Step 3: If $L_c \leq MT_i$ then $T_{ci}^* = MT_i$ and $MT_i = MT_i - T_{ci}^*$ and go to step 4.

Step 4: If $MT_z = 0$ for all $z \in M$ or $c = n$ then go to step 5. Otherwise, let $c = c + 1$ and loop $= c$ and return go to step 2.

Step 5: Calculate $Z^* = \sum_{c=1}^{n} \mu_c T_{ci}^*$ and stop.

Output: The best solution of total effectiveness is Z^* and T_{ci}^* is the global maximize point of problem used by FA and DA.

In the experimental tests, for example, we solve the effectiveness problem with the number of 7 cases for experimental parameters and the number of justice team for experimental parameters was 3. The total time available (deadline) for justice decision is $MT = 90$. Therefore, we generated parameters μ_i, U_i, and L_i by following two problem with a uniform distribution e.t. in the form $P1$: $\mu_i = (1,90), L_i = (1,30), U_i = (1,90)$ and $P2$: $\mu_i = (1,90), L_i = (1,60), U_i = (1,90)$. Table 1 shows the case study using $P1$ the data set into part 1 and the total effectiveness into part 2, respectively.

Table 1. The case study

Example data set							
c	1	2	3	4	5	6	7
i	1	2	3	-	-	-	-
μ_c	27	25	20	15	13	10	8
L_c	2	6	13	28	16	29	17
U_c	85	61	29	34	59	32	48
Calculate Z^*							
c	1	2	3	4	5	6	7
i	2	1	3	1	3	2	1
T_{ci}	71	33	15	87	49	30	9

Table 2. The results of FA and DA for the problem

P	$(n:m)$	DA Effectiveness % Deviation Avg.	DA Time (s)	FA Effectiveness % Deviation Avg.	FA Time (s)
$P1$:	(50 : 3)	0.01	603.81	0.08	733.30
	(50 : 4)	0.02	716.05	0.04	862.47
	(50 : 5)	0.01	833.30	0.02	918.17
$P2$:	(50 : 3)	0.05	715.97	0.13	752.46
	(50 : 4)	0.09	828.45	0.14	853.54
	(50 : 5)	0.01	952.46	0.02	976.25

The assignments described above by executing step 2 – 5, step 5 : Calculate $Z^* = (27\times71) + (25\times33) + (20\times15) + (15\times87) + (13\times49) + (10\times30) + (8\times9) = 5356$ and stop. Next, the output of total effectiveness $Z^* = 5356$ is the best solution found by FA or DA.

3. Results and Discussion

The computational and effectiveness results are generated using FA and DA. Table 2 shows the total available CPU time (seconds) in finding an optimal solution where 25 problem instances were generated. The number of cases for experimental parameters was 25 with the number of justice team for experimental parameters was 3, 4 and 5, Percent deviation is 100*[Optimal solution - FA or DA solution]/[Optimal solution]. For 25 instances, the average of percent deviation is $\sum_{i=1}^{25}$ [percent deviation]/25. Experimental results show in Table 2 articulate that time of optimal solutions and average of percent deviation for maximize effectiveness of DA are better than those of fast FA. In the results, it can be seen that when the number of cases was 50 and the number of justice team was 3 for experimental parameters, P1(50:3,4,5) the average of percent deviation of DA is 0.01, 0.02, and 0.01 and total available time of DA is 603.81, 716.05, and 833.30, respectively. It can be seen that the result is better than the fast FA. Furthermore, in the P2(50:3,4,5) the average of percent deviation of DA is 0.05, 0.09, and 0.01 and total available time of DA is 715.97, 828.45, and 952.46 respectively. The result is also better than fast FA in measuring performance.

4. Conclusion

In this paper, the proposal to increase effectiveness of court case assignment through a MILP is presented. A metaheuristic algorithm is used to find the optimal solution for assignment problem. Experimental results show that time of optimal solutions and average of percent deviation for maximize of DA are better than those of fast FA. The performance of DA is better than that of FA for maximizing effectiveness of case assignment in court of justice.

In the future work, it will be useful to improve metaheuristic optimization for the fast computing and defining parameter optimization for the assignment problem in legal fields.

Acknowledgments

This research is supported by Research and Technology Transfer Affairs (KKU-NKC60-009). The authors would like to also thank to Faculty of Applied Science and Engineering and Faculty of Integrated Social Sciences Khon Kaen University, Nong Khai Campus for funding this research.

References

[1] Stephen L, Deborah I, and Philip B 2011 Unintended Consequences: Constraint of Customary Justice in Post-Conflict Liberia in Deborah H. Isser, ed. *Customary Justice and The Rule of Law in War-Torn Societies* United States Inst. of Peace.

[2] Phitayaporn T 2004 Strengthening the Independence and Efficiency of the Judiciary in Thailand *Chulalongkorn University Law Journal* pp 45-53.

[3] Takeshi K 2015 Mixed integer programming approach on examination proctor assignment problem *Procedia Computer Science* vol 60 pp 818 – 823.

[4] Julia R, Jürgen Z, and Thorsten G 2012 Mixed-integer linear programming for resource leveling problems *European Journal of Operational Research* vol 221 pp 27–37.

[5] David B, Kjetil H, Asmund O, and Jan N 2015 A mixed integer linear programming model applied in barge planning for Omya *Journal of Operations Research Perspectives* vol 2 pp 150–155.

[6] Christodoulos F 2005 Mixed Integer Linear Programming in Process Scheduling: Modeling Algorithms and Applications *Journal of Annals of Operations Research* vol 139 pp 131–162.

[7] Charbel R and Jean A 2012 ACO for the Surgical Cases Assignment Problem, *Journal of Medical System* vol 36 pp 1891–1899.

[8] Anil G, Shiv J, and Surbhi G 2013 A Heuristic Method of Cell to Switch Assignment in Mobile Communication Networks Using Firefly Algorithm, *International Journal of Scientific & Engineering Research* vol 4 no 7 pp 1971-1977.

[9] Yang X 2008 *Nature-Inspired Metaheuristic Algorithms* (Luniver Press).

[10] Seyedali M, 2016 Dragonfly algorithm: a new meta-heuristic optimization technique for solving single-objective discrete and multi-objective problems *Neural Computing and Applications* vol 27 pp 1053 - 1073.

Research on the Influence and Correction Algorithm of Illuminance Intensity of Photoelectronic Imaging Bridge Deflection Measurement Accuracy

Yishu Zhou[1,a], Jinjin Cao[1,b], Yilu Zhang [2], Huayang He[1] and Jinning Zhang[1]

[1]Research Institute of Highway, Ministry of Transport, Beijing, China
[2]Beijing University of Posts and Telecommunications, Beijing, China

[a] ys.zhou@rioh.cn; [b] injincao7210@126.com

Abstract: As a non-contacted non-destructive testing equipment, the bridge deflection measurement instrument based on the principle of photo-electric imaging has the characteristics of high measurement precision, strong dynamic and static deflection observation ability and long observation range, and has a vast application prospect in the field of bridge structural health monitoring. The photoelectronic imaging bridge deflection instrument obtains the deflection measurement result based on the target image, and the image quality is an important influential factor of the measuring accuracy. Illuminance intensity is an important factor affecting the image quality, and will have an important influence on deflection measurement in different weather conditions and different time periods. This paper studies the effect of the different illuminance intensity conditions on the accuracy of the photoelectronic imaging bridge deflection instrument, then a correction method of deflection measurement accuracy based on illuminance intensity measurement is proposed, and the brightness balance correction to the experimental pictures is refined.
Keywords: Photoelectric Imaging, Illuminance Intensity, Brightness Balance Correction, Bridge Deflection

1. Introduction

Bridges are an important component of modern transportation systems, and load testing and health monitoring are essential for them. As one of the most important bridge structure parameters, beam deflection[1,2] reflects directly the vertical rigidity of the bridge structure, and is closely related to the bearing capacity of a bridge and its ability of resisting dynamic load such as from earthquakes. The common measuring tools of deflection are theodolite, level, dialgauge, etc. but the limitations of manual measuring have greatly increased the difficulty of the site work. At the same time, some automatic deflection measurement methods[3], such as the method of total station, GPS observation method, inclinometer method and special deflection method (photoelectric bridge deflection instrument, laser deflection instrument, collimation laser projection type deflection monitoring system and microwave Interferometric Radar deformable measurement system), etc. are used in bridge deflection monitoring work. Among them, the photoelectric bridge deflection instrument is widely used in bridge deflection detection [4]because of its relatively high observation accuracy and simple engineering operation.

2. System Composition and Detection Principle of Photoelectric Bridge Deflection Instrument

The diagram of the structure and principle of the photoelectric bridge deflection instrument[5] is shown in figure 1. It is mainly composed of three parts, which are the luminescent target, the photoelectric digital camera and the computer acquisition and processing system. The luminescent target is fixed on the point of measurement on the bridge body. The photoelectric digital camera is fixed in the stable mechanical device. The luminescent target is imaged on the photosensitive surface of the digital camera by the optical system of the camera. The computer collects the image of the luminescent target from the camera by the image acquisition card, and the exact location of the image point is determined by image processing recognition. Finally, the relationship between the object determines the location of the cursor target. When the bridge deflection is changed, the light target also produces the equivalent displacement with the bridge, thus the change of the deflection of the bridge is determined.

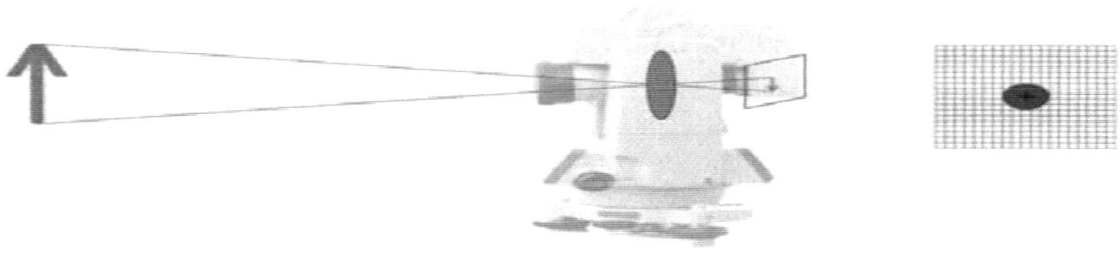

(a) Schematic diagram of the structure of a photoelectric bridge deflection instrument

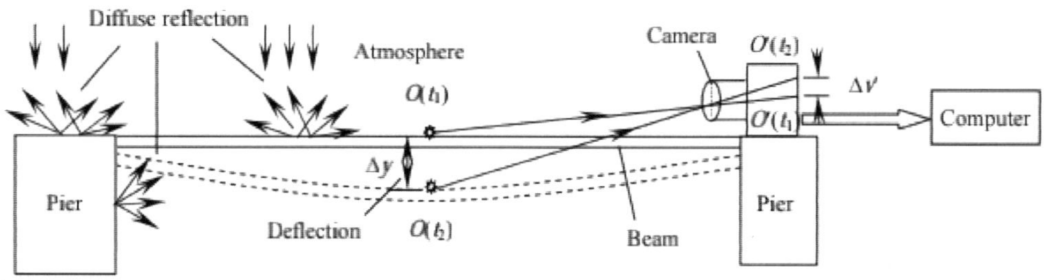

(b) Schematic diagram of imaging theory

Figure 1. Schematic diagram of the structure and imaging of a photoelectric bridge deflection instrument system.

3. Environmental factors affecting the accuracy of measurement

When photoelectric bridge deflection instrument is working in outdoor environment, the work is affected by many factor. The huge impact of light, rain and fog, atmospheric disturbances and other environmental factors on the measurement system can not be ignored. As shown in figure 2, the intensity of light will lead to the reduction of the contrast between the luminescent target and the background light, and it will be impossible to take a target image with a clear outline; the presence of rain, fog, haze and other particles will cause a decline in the visibility of the surrounding environment, making it impossible for the camera to shoot a clear target image. Based the accuracy of the result of bridge deflection measurement based on image processing technology will inevitably be affected by the imaging environment. Among many factors, the illumination has a more serious effect on the accuracy of deflection measurement. Whether the brightness is too dark or too bright, it may result in the target image may not be able to capture the location information.

Therefor, this paper conducts a brightness test experiment focused on the influence of the light intensity on the brightness of the picture taken by the detector, and the picture's brightness balance correction, which will is an important pretreatment process to improve the measurement accuracy.

Figure 2. The environmental factors that affect the accuracy of measurement

4. Experimental Process and Result Analysis

The experiment is taken outside. In a wide open field, a fixed building is selected as reference, and the target is erected before the reference material to keep the lights on. In the experiment, HPQN-X photoelectric bridge deflection instrument is used for shooting, as shown in figure 3. The illuminance meter is used to detect the illuminance in real time. A total of 10 pictures of the building under different illuminance are taken, and as shown in figure 4, the brightness of these pictures is analyzed.

Figure 3. Photoelectric bridge deflection instrument used in the experiment

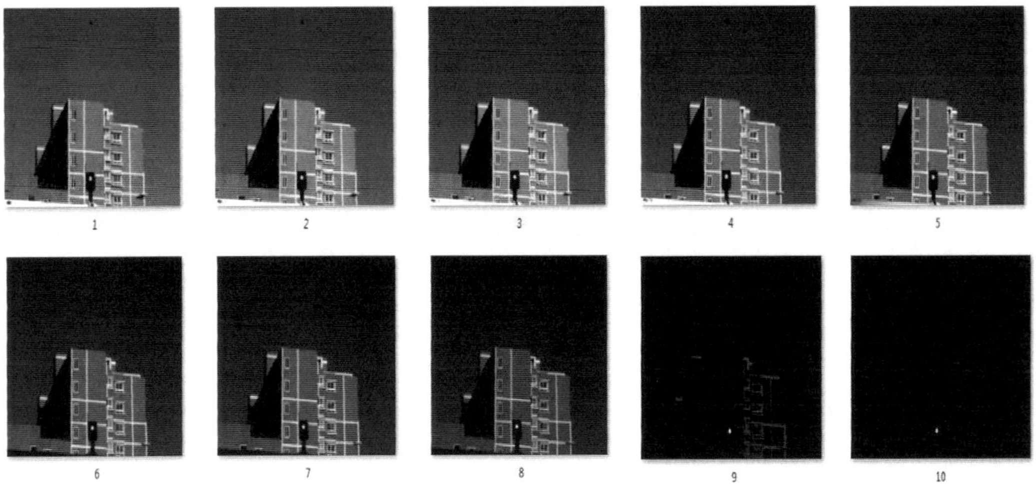

Figure 4. 10 pictures under different illuminance intensity

Each picture is analyzed at 10×10 pixel, and the resulting luminance is shown in table 1:

Table 1. 10×10 pixel luminance value for each picture

Picture	1	2	3	4	5
Luminance value	0.1882	0.1647	0.1176	0.0980	0.0824
Picture	6	7	8	9	10
Luminance value	0.0627	0.0510	0.0353	0	0

Using the illuminometer for real-time measurement, the ambient illuminance intensity at the shooting time of the 10 photos are shown in table 2:

Table 2. Background light of photo shooting

Picture	1	2	3	4	5
Real time illumination/$\times10^4$ Lux	5.72	4.95	3.81	3.36	2.20
Picture	6	7	8	9	10
Real time illumination/$\times10^4$ Lux	1.80	0.82	0.55	0.31	0.14

The analysis shows that with the decrease of ambient illuminance intensity, the brightness value of the picture gradually becomes smaller. When the illumination intensity drops to 0.3 $\times10^4$Lux, the brightness value of the picture of the selected region is 0. Therefore, under the uniform illumination condition, with the decrease of the illumination intensity, the image quality of the picture drops; by contrasting the grayscale of the 10 pictures it can be known that when the target brightness is constant, the intensity of the background illumination decreases, and the contrast with the target brightness is enhanced, which will make the target prominent and easy to be captured by camera. Therefore, in the actual detection, the target and background light with strong contrast are more conducive to the accuracy of bridge deflection detection.

5. Luminance Balance Correction

On the basis of THE luminance analysis, this paper further improves the brightness balance of images by digital image processing technology[6], in the hope of achieving similar brightness by luminance equalization[7,8].

Dividing the image into several small blocks, the luminance information of each sub-block is calculated or estimated, and the brightness of each sub-block is adjusted so as to make the image's brightness of the uneven illumination uniform. The difficulty of the brightness equalization algorithm lies in the extraction of brightness, the RGB color space does not give the brightness information of the color directly, but the calculation formula of L used when RGB is converted to HSL model.

$$L = \frac{(\mathrm{maxcolor} + \mathrm{mincolor})}{2} \tag{1}$$

Among them, maxcolor and mincolor represent the maximum and minimum value of R, G and B respectively in the picture.

From Equation 1, the brightness of the picture is related to the maximum and the minimum total

IOP Conf. Series: Journal of Physics: Conf. Series **1061** (2018) 012006 doi:10.1088/1742-6596/1061/1/012006

color component of the image, so the RGB three-channel sum can simulate the brightness information of the image. For the convenience of calculation, the image is converted into a grayscale image by rgb2gray conversion, the maximum brightness value in the grayscale image is selected as the standard value, the pixel brightness of all the images is adjusted according to the value, and the brightness of the image is adjusted to achieve the current best condition.
The algorithm principle is as follows:

the picture is converted to grayscale, the average brightness of each picture is calculated, assuming the picture is M × N pixel size, the grayscale is 0 to L, then the average brightness of the picture is

$$L_{avg} = \frac{\sum_{i=0}^{M=L}\sum_{j=0}^{N=L} q(i,j)}{M \times N} \tag{2}$$

Where q (i, j) is the brightness value of the pixel point (i, j) in the image.

Taking the maximum brightness value as the standard brightness value, the ratio of the standard brightness to the brightness of each image is calculated

$$L' = \frac{L_s}{L_{avg_k}} \tag{3}$$

Where L_{avg_k} is the average brightness of the Kth image.

Taking the ratio of the Equation 3 as the image brightness equalization coefficient, the brightness of each picture is adjusted so that all images are output to the same brightness range to achieve the brightness equalization.

The 8 pictures of Figure 4 are selected for brightness adjustment, and the balance effect is shown in Figure 5.

From Figure 4, it can be known that the brightness of the untreated picture changes with the illumination intensity, and Figure 5 is aboutt the treatment of the picture through the brightness balance correction, which makes the picture under different illuminance reach the same brightness effect. The effect of background illumination on picture brightness is eliminated. Because the brightness values of the last two pictures in Figure 4 are too low, the brightness balance is invalid when the reference is not clear.

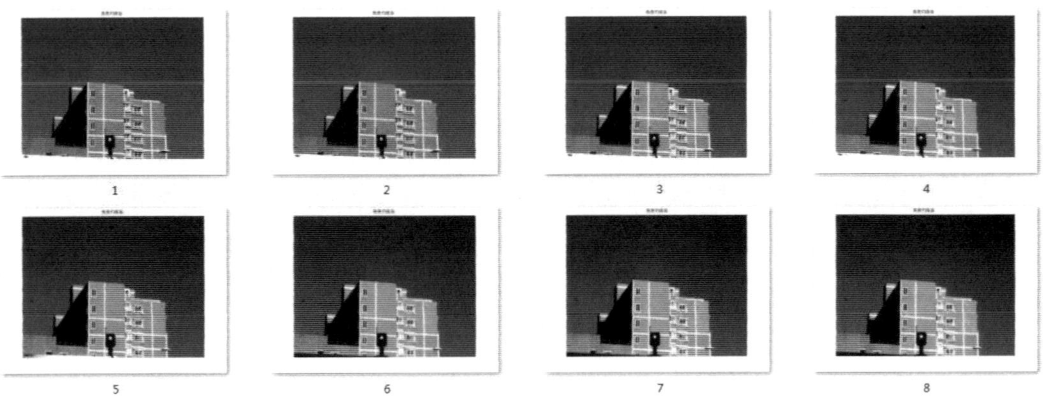

Figure 5. The result of the brightness equilibrium.

6. Conclusion

The paper analyzes the influence of illuminance intensity on the pictures taken by photoelectric bridge deflectometer, and the effect of light on the image quality is analyzed emphatically. The brightness balance of the pictures under the different illumination of the actual test is corrected. The experimental results show that illuminance intensity has a significant impact on the photoelectric imaging, and the brightness balance can improve the image quality. The brightness equalization can be regarded as an important pretreatment step of the photoelectric bridge deflection instrument, thus further improving the deflection measurement precision.

References

[1] Kuo-Wei Liao,Min-Yuan Cheng,Yung-Fang Chiu,Jin-Han Lee. Preliminary bridge health evaluation using the pier vibration frequency[J]. Construction and Building Materials,2016,102.
[2] J.M. Ko,Y.Q. Ni. Technology developments in structural health monitoring of large-scale bridges[J]. Engineering Structures,2005,27(12).
[3] Gorazd Lipnik, M. Sc. E., Kovacic Bostjan, Vojko Matko.Computer based dynamic measurements of displacements on bridges[C].First ISA/IEEE Sensors for Industry Conference (SIcon-01).2001.142-144.
[4] A. Cunha,E. Caetano,F, et al.Recent perspectives in dynamic testing and monitoring of bridges[J].Structural control and health monitoring,2013,(6):853-877.
[5] Jie Fang;Wenwen Zhang;Weiji He;Qian Chen;Guohua Gu.Evaluation of the Photoelectric Performance Parameters Measurement for Electron Multiplying CCD.Proceedings of OTA 2016—conf.5 optical measurement technology and instrument.2016.
[6] Ying XIN,Wei XUE.Application of some Valid Methods in Logs Counting System based on Digital Image Processing[C].2010 IEEE 17th International Conference on Industrial Engineering and Engineering Management.2010:400-403.
[7] Lim, Sheng Hoong,Isa, Nor Ashidi Mat,Ooi, Chen Hee, et al.A new histogram equalization method for digital image enhancement and brightness preservation[J].Signal, image and video processing,2015,(3):675-689.
[8] Thien Huynh-The; Ba-Vui Le; Sungyoung Lee; Thuong Le-Tien; Yongik Yoon.Using weighted dynamic range for histogram equalization to improve the image contrast.EURASIP Journal on Image and Video Processing.2014,Vol.2014 (1),pp.1-17.

Samawa Language: Part of Speech Tagset and Tagged Corpus for NLP Resources

Trienani Hariyanti[1,2], Saori Aida[1] and Hiroyuki Kameda[1]

[1] School of Computer Science, Tokyo University of Technology, 1404-1 Katakuramachi, Hachioji, Tokyo
[2] Department of Computer Science, Sumbawa University of Technology, Batu Alang, Sumbawa, Indonesia

E-mail: trienanihariyanti@gmail.com

Abstract. Samawa language is one of the local languages in Sumbawa Island, Indonesia. It was categorized as an under-resourced language since there is only a small number of books written and studies reported in the literature. The availability of Samawa corpus is an alternative to preserve Samawa as a cultural heritage. In this work, we describe our first effort to build the first Samawa corpus and assign 24 part of speech information to each token based on lexical and grammatical rules of Samawa. The raw data used was collected from the manuscripts, textbooks, magazines, and text from websites. It was cleaned and normalized by UTF-8 Unicode standard scheme and converted into XML format with TEI guidelines. The result is a Samawa tagged corpus of 739 sentences that contain 11,799 tokens and can be used for developing tools in many NLP applications.

1. Introduction

Samawa language is one of the local languages in Sumbawa Island, Indonesia. It is used as the second language in daily communication besides Bahasa Indonesia as the official language in most Indonesian areas. Also, it was spoken by around 500,000 speakers spread in average from western (Seteluk sub-district) to the east (Empang sub-district) in Sumbawa Island. Despite the relatively large a number of speakers, Samawa was categorized as an under-resourced language where there is only small number of books in Samawa published to date, and studies have been reported in the literature. Moreover, in recent year, the subject of Samawa does not exist in curriculum's policy in the school. It may have severe consequences for the existence and sustainability of Samawa for the next generation in the future. The availability of corpus might be an alternative to preserve Samawa as a cultural heritage and making it more accessible. A corpus represents a documentation of language in use and provides linguistic diversity. It can be useful for developing many tools in Natural Language Processing (NLP). Therefore, we initiate a new NLP research to build a Samawa corpus with its part of speech tagset.

The raw data used in Samawa corpus was collected from the manuscripts, textbooks, magazines, and text from websites. Then in the preprocessing phase, it was cleaned and normalized by Unicode in plaintext format. Regarding standard of the text corpus in digital form, the plaintext was converted to XML (eXtensible Markup Language) appropriated with Text Encoding Initiative (TEI) guidelines [1]. We also added 24 part of speech information in the corpus which consists of 11,799 tokens arranged into 739 sentences.

The rest of this paper is organized as follows. In section 2 we describe related work regarded to morphology and grammatical rules of Samawa. We generally present Samawa language and its morphology, and introduce our methods to create the corpus from preprocessing the raw text to manually tagging in section 3 and 4, respectively. In the last section, we summarise our conclusions and present future works.

2. Related Work

Several studies have already been done related to Samawa language. In 1986, Sumarsono et al. [2] proposed a general phonology, morphology of Samawa such as affixation, reduplication, composition and grammatical structure of Samawa language. Besides, they presented kind, function and the meaning of morphemes. In other research, Seken et al. in 1990 focused on Samawa morphology. They studied some of the rules that have not been described in Sumarsono [3]. In the same year, Mahsun [4] corrected the lack of previous studies in the form of separation and grouping of some affixes based on their correlation and meaning in Jereweh dialect. Mahsun also proposed a morphology of Jereweh dialect in [5] and mapped the dialects of Samawa in diachronic dialectological studies [6]. Then, Kasman in 2016 studied about the position and function about lingual form /q/, /n/, /èn/, /èng/, and /s/ in Tongo dialect [7]. The last, Asmadi [8] provided morphology process in Samawa based on descriptive approach.

3. Samawa Language and its morphology

Samawa language has four dialects, i.e. Taliwang, Jereweh, Tongo, and Sumbawa Besar. One of the most critical linguistic characteristics of four dialects is the realisation of the high, front of the ultima syllabic vocals which end with the dorso-velar consonant, i.e., /s/, /t/ and /n/ [9]. Sumbawa Besar dialect was chosen as the standard of Samawa language because it has a relatively wide spread of speakers. On the other hand, the difference between each dialect is not too noticeable, lies only in the lexical and grammatical sides in some affixes [5].

Phonology system of Samawa language divided into vowels and consonants. Vowels consists of 8 phonemes such as /i/, /e/, /ɛ/, /ɔ/, /o/, /ð/, /u/, and /a/. Consonants consists of 19 phonemes such as /p/, /b/, /t/, /d/, /c/, /j/, /k/, /q/, /s/, /h/, /m/, /n/, /ŋ/, /ñ/, /r/, /l/, /w/, and /y/.

Samawa language has three kinds of affixation, i.e., prefix, confix, and combination of them. Affixation means a morphology process of adding affixes to the base form of words. It can change the form, part of speech and the meaning of a word. Table 1 shows kinds of affixes in Samawa language.

Table 1. Affixes in Samawa language

Prefix	Confix	Combination of Affix
ba-	bakaN-	baka-
gaN-		basaN-
ka-		YagaN-
kaN-		yaka-
N-		yapa-
pa-		yasaN-
paN-		
ra-		
raN-		
sa-		
saN-		
ya-		

IOP Conf. Series: Journal of Physics: Conf. Series **1061** (2018) 012007 doi:10.1088/1742-6596/1061/1/012007

4. Samawa Corpus Creation

Building Samawa corpus is a quite challenging task. First, the amount of document and textbook in Samawa language is strongly limited in size and variation since the speakers only used the Samawa language in daily communication and cultural event. Second, almost documents are particularly difficult to handle. So we have to type them manually. In the following subsection, we describe each phase of building Samawa corpus we made.

4.1. Data Collection

Raw data for corpus was collected from Sumbawa humanists, archive office of Sumbawa and language office centre of West Nusa Tenggara from the various time that the data has published. It is also essential that the data must be compiled from various domain or genre in order to represent a corpus with language variety. Table 2 shows the frequency distribution of genre in our collection.

Table 2. Frequency distribution of genre.

Genre	Words (%)
History	4.66
Culture	3.98
Relationship	4.37
Religion	25.20
General	61.79
Total	100

4.2. Preprocessing

There is one main process in the preprocessing phase, i.e. text cleaning. In text cleaning process, the data was standardized into Unicode character encoding scheme of UTF-8. This encoding standard was chosen because of its flexibility and portability across platform [10]. Since some of the documents were written using original phonemes that follow phonology system above, then we normalized them with existing Latin characters. For instance, a word "**Tongtongè**", we normalized the text to become "**Tongtonge**". Later, we corrected some spelling mistakes based on Samawa dictionary in [11, 12] and deleted white spaces.

4.3. Part of Speech Tagset

Recently, there are many lists part of speech or tagset for English. Penn Treebank, Brown and CLAWS are commonly used for English tagset. A Tagset for language processing gives information about a word and its neighbours. It can be useful to build a language model [13].

Sumarsono et al. [2] classified Samawa tagset into 11 categories. These are noun, pronoun, verb, adjective, particle, number, question, coordinate conjunction, negate and modal. We used it as the base of our design to define our original tagset. Then, we analyzed the needs tagset of Samawa and proposed modification tagsets in Table 3.

Table 3. Our original Samawa tagset.

No.	Tagset	Description	Example
1	CC	Coordinate Conjunction	atawa, tapi, ke
2	CD	Cardinal Number	telu, satenga, saratis, peno
3	DT	Determiner	Datu, Dea, Lalu, Lala

4	FW	Foreign Word	Pernyataan, sobat
5	IN	Preposition	pang, kaling, muntu, tawa
6	JJ	Adjective	pisak, le, ngamuk, belo
7	MD	Modal / Auxiliary	bau, ya, perlu, musti, ka
8	NEG	Negate	no, siong, nopoka, na
9	NN	Noun	to, nawar, tone, sampi
10	NNP	Proper Noun	Indonesia, Lebaran, Jemat
11	NND	Measure Word	lolo, kg, lamar
12	OD	Ordinal Number	kalima, mula
13	PR	Demonstrative	deto, deta, ta, nan, ana
14	PRP	Personal Pronoun	kaji, kelam, nene, nya, tu
15	RB	Adverb	paleng, majan, mala, bae
16	RP	Particle	ne, mo, si, ke, po
17	SC	Subordinate Conjunction	lamen, ande, lebe, kele
18	SYM	Symbol	+, -, ß, ±, #, $, ¥, @, (), []
19	UH	Interjection	aida, aina, ba, waida, waina
20	VB	Verb	kumir, labir, kuris
21	VBT	Transitive Verb	nali, meli, sabongkang
22	VBI	Intransitive Verb	manto, belangan
23	WH	Question	kuda, pidan, meluk, pida
24	Z	Punctuation	,;”...:

4.4. Manually Tagging of Samawa Corpus

In this subsection, we manually assigned a tagset of 24 tags to Samawa corpus based on our principle guidelines, i.e., three books of Samawa morphological and grammatical rules [2, 3, 4]. The process of annotating the corpus was divided into two major steps. First, we manually tagged 100 sentences using part of speech tagset to each token with a slash character. Each token is tagged mainly by checking its grammatical category to see if it is either word or punctuation. During this manual tagging process, we have noted the primary issue that the token have more than one tag (ambiguous). We listed, for disambiguation every rules that can be useful for tagging the next sentences with the same pattern. Moreover, it also useful as grammatical rules to develop rule-based tagger in the future.

Second, we defined and implemented a hierarchy structure of the data files of Samawa corpus based on Text Encoding Initiative (TEI) guidelines for XML. It was inspired by previous work of XML based research on Myanmar National Corpus [14]. A simple program was developed to convert

IOP Conf. Series: Journal of Physics: Conf. Series **1061** (2018) 012007 doi:10.1088/1742-6596/1061/1/012007

the result of manually tagging in XML. The hierarchy structure of Samawa corpus file can be seen in Figure 1.

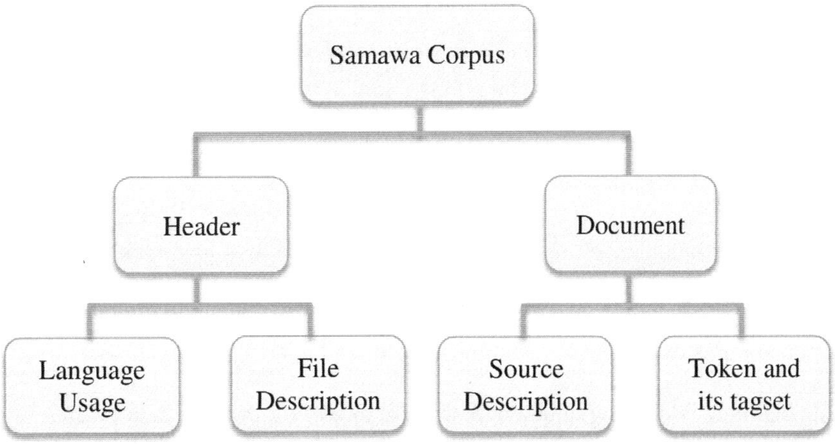

Figure 1. Hierarchy structure of Samawa corpus file.

Furthermore, we used the script by Jacob Perkins in [15] to see the lexical diversity of our corpus. There are 11,799 words in total and 1,674 unique words. Figure 2 shows the distribution of 10 major tags mostly used in Samawa corpus.

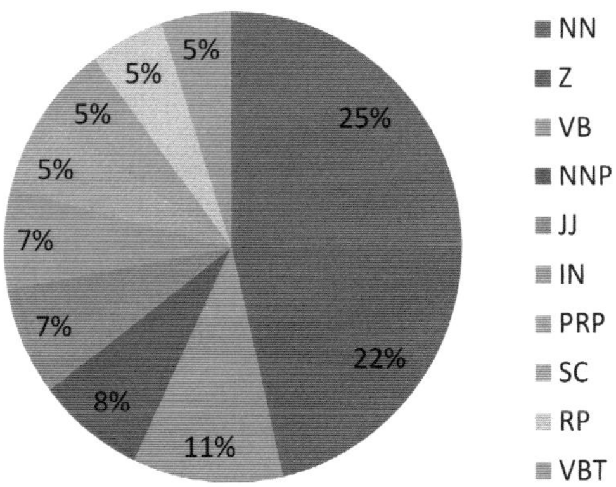

Figure 2. Part of speech tags coverage in Samawa corpus.

5. Conclusion and Future Works

This work describes our initial effort to build a new Samawa corpus. It is available in plaintext and XML format with TEI standard. We designed a tagset of 24 tags that contains the information about part of speech. We noted some grammatical rules which could be used to speeding up our manually tagging. There are, however, some limitations in our corpus such as the absence of some domain of text in Samawa language and its size of data. We wish to extend our tagged corpus and its variation using automatic tagger by collaborating with linguistics that leads to developing many tools in NLP with the corpus.

Acknowledgments
This work is supported by Tokyo University of Technology, Japan. Mrs Trienani Hariyanti would also thank to the Indonesian Endowment Fund for Education (LPDP) for gratefully acknowledge funding on her study.

References
[1] Consortium T E I 2008 *TEI P5: Guidelines for electronic text encoding and interchange* (URL: http://www. tei-c. org/release/doc/tei-p5-doc/en/Guidelines. pdf [September, 2011])
[2] Sumarsono, Nadera M, Basuki S, Merdhana N and Mertha N 1986 *Morfologi dan Sintaksis Bahasa Sumbawa* (Pusat Pembinaan dan Pengembangan Bahasa, Departemen Pendidikan dan Kebudayaan)
[3] Seken I K, Basuki S, Gosong I M and Putu Antara I G 1990 *Morfologi Bahasa Sumbawa* (Pusat Pembinaan dan Pengembangan Bahasa, Departemen Pendidikan dan Kebudayaan)
[4] Mahsun 1990 *Morfologi Bahasa Sumbawa Dialek Jereweh* (Yogyakarta: Universitas Gadjah Mada)
[5] Mahsun 2006 *Kajian Morfologis Bahasa Sumbawa Dialek Jereweh* (Gama Media)
[6] Mahsun 1995 *Dialektologi Diakronis - Sebuah Pengantar* (Gadjah Mada University Press)
[7] Kasman N F N 2017 Kedudukan dan Fungsi Satuan Lingual/q/,/n/,/èn/, /èng/, dan /s/ dalam Bahasa Samawa Dialek Tongo Subdialek Lebangkar *Kandai* **12** 241–254
[8] Asmadi 2016 Proses Morfologis pada Kultur Bahasa Etnis Sumbawa *RETORIKA: Jurnal Ilmu Bahasa* **2** 214–232
[9] Mahsun 1994 *Penelitian Dialek geografis bahasa Sumbawa* PhD Thesis (Universitas Gadjah Mada)
[10] Chungku C, Rabgay J and Choejey P 2011 Dzongkha Text Corpus *Conference on Human Language Technology for Development* pp 34–38
[11] Muslim N, Nuryati, Hartini and Yudiastitni N M 2016 *Kamus Samawa - Indonesia* (Kantor Bahasa NTB)
[12] Amin U and H.M. A H 2016 *Kamus Bahasa Sumbawa - Indonesia* (Lembaga Adat Tana Samawa (LATS): Paboat Adat Bidang Bahasa dan Sastra Daerah)
[13] Jurafsky D and Martin J H 2000 *Speech and language processing: an introduction to natural language processing, computational linguistics, and speech recognition* (Upper Saddle River, N.J: Prentice Hall)
[14] Ko W K and Phyo T Z 2008 Selection of XML tag set for Myanmar National Corpus *IJCNLP* pp 33–40
[15] Perkins J 2014 *Python 3 text processing with NLTK 3 cookbook: over 80 practical recipes on natural language processing techniques using Python's NLTK 3.0* (Birmingham: Packt Publ)

IOP Conf. Series: Journal of Physics: Conf. Series **1061** (2018) 012008 doi:10.1088/1742-6596/1061/1/012008

Probabilistic Implicit Multitask Feature Learning Based on Weight Matrix Partition

Zhang Yu[1], Xin Zuo[1] and Jian-wei Liu[1]

[1]Department of Automation, College of Geophysics and Information Engineering, China University of Petroleum , Beijing (CUP), Beijing, 102249, China

*[1]zhangyunepu@163.com; zxin@cup.edu.cn; liujw@cup.edu.cn

Abstract. Multitask learning (MTL) can be considered at two levels, considering the multiple features attribute that can be migrated between the most tasks simultaneously called MTL on task level, and only considering the features attribute that can be migrated between a very few tasks called MTL on feature level. A single feature is composed with an overlying structure including these above two levels, since the similarity between the multi-tasks is not unique. In the previous MTL studies, in task level, the shared feature structure of inter-task is the leading approach, and in feature level, only considering a partial of salient features is more prevalent. However, this model and structure assumptions are violated on most complex learning scenarios, and some specific feature structures is also limited. In this paper, from the Bayesian viewpoint, we propose a new method, the weight matrix is defined by stochastic variables which probabilistic distribution is overlaid, the relationship in task level are no longer represented by a unified structure ,and relationship in feature level are not limited to concrete structural forms. The overlying structure is defined by two parts: multidimensional Gaussian mixture distribution according to the overall features to describe the task relationship and one-dimensional Gaussian mixture distribution to describe the feature relationship on every feature. These two parts can be elaborated by the information contained in each Gaussian sub-classes separately, then we determine the closeness of the relationship by calculating the posterior probability distribution. The advantage of our method is that we can operate the weight matrix more flexible and make full use of the probability distribution to get abundant expression in the task level and grasp the key factors in the feature learning process.

1. Introduction

The concept of multitasking learning was first proposed in [1], the key of the MTL is to exploit the multitask relationship and seek for valuable information from similar tasks to improve the learning efficiency. When the sample size in some tasks are not sufficient, single task learning will be rank deficient and is easy to appear over fitting, in order to alleviate these problems, the consideration about characteristics migration of the task is necessary, one can achieve the purpose of expanding the samples indirectly. In the early work by [2,3],they presented a model that all related tasks share a common structure and uniform feature correlation is assumed, and achieved the row sparse effect to control the number of features shared across the tasks, the result is only a few features are utilized by other tasks, and the other useful information is lost. This can cause negative migration on some tasks which have sufficient samples, and very few task sets are completely similar in feature structure.

To solve this problem, some researchers [4, 5] design classification according to the overall similarity of features, the tasks is divided into the most relevant task groups, and no information exchange between different groups. However, this method is still based on the hypothesis of task level that always concern the majority features, and it will meet some limitations when application. Such as recommendation systems, there are significant differences in their respective tasks and large scale features transfer are almost not exist, impact of salient features existing in some tasks will lost in the task level learning process. So some scholars are devoted to multitask feature learning and there appear some methods considering the characteristic of the feature individually. [6, 7] proposed a superposition models called robust MTL contain shared and outlier features. Then [8] apply "dirty" model to turn the outlier features into sparse structure.

For exploring the sparse characteristics of these special features, [9] established a method to cluster the features in the same dimension of each task, they point out these specific features have a remarkable promotion of learning effect. Recently, other researchers [10, 11] have proposed methods based on a partial special feature set, organized the special features into feature blocks and there is no connection between different feature blocks. But in MTL, the relationship between the features may be across tasks and across dimensions, this can be explained by "the eavesdropping mechanism" in [1], the connection between these features makes it possible to benefit from other tasks, and basically, this relation is embodied by linear relationship and is not limited only in the same dimension as feature block, because we know that in single task learning, there are coupling information among features, Therefore in MTL, this effect will not only be eliminated, but can be passed through different dimensions between tasks(see Figure 1). Considering some original related features are always divided into different sub blocks in [10], the relationship between these sub blocks shouldn't be ignored.

Inspired by the idea, we put forward a new probabilistic method of multitask feature learning called MTPFL, the model is a superposition form which can make clustering in task level and feature level simultaneously. In task level, we divide tasks into the most relevant classes based on overall features. And in feature level, we integrate the sparse features except for the shared structure, looking for connections between them by making clustering. It is no longer limited to form blocks in feature level, moreover, we establish connections between whatever the same or different dimensions.

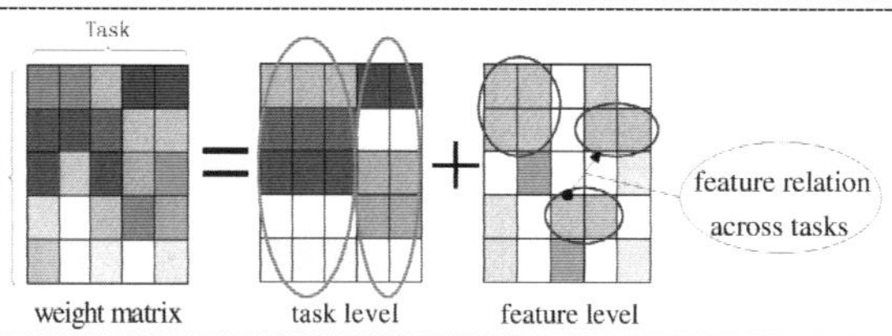

Figure 1. A general structure of MTPFL, the correlation on tasks and features are elaborate, there exist some discrete correlation between features.

Multitask information migration is reflected by the weight matrix, another major advantage of our method is the flexibility of the representation of the weight matrix W. The usual process calculating the value of the W is a continuous iterative optimal algorithm, which under the constraint of the objective function and the structure of the MTL is prescribed. In this article, we do not specify such structural constraints but construct a generative model using Gaussian mixed model (GMM) for the two levels' relation mining in W, the Gaussian mixture distribution has good performance which can fit any probabilistic distribution, so the description of relationship is abundant and complete, we only need to utilize the characteristics of the dataset itself to find useful information.

2. Model Formulation

In this section, we will derive our proposed probabilistic model on multitask learning. Supposing there are totally M tasks, each task has a sample set $D_m = \{x_i^m, y_i^m\}_{i=1}^{N_m}$, where $m = 1, \ldots, M$. To learn the model, we formulate the linear model on input-output pairs:

$$\begin{aligned} y_i^m &= w_0^m + x_{i,1}^m w_1^m + x_{i,1}^m w_2^m + \ldots + x_{i,D}^m w_D^m + \varepsilon_m = w_0^m + (x_i^m)^T W^m + \varepsilon_m \\ &= w_0^m + (x_i^m)^T (U^m + V^m) + \varepsilon_m \quad i = 1, 2, \ldots, N_m \end{aligned} \tag{1}$$

Where $x_i^m \in R^D$ is the input vector of the m-th task, $X^m = [x_1^m, x_2^m, \ldots, x_{N_m}^m]^T$ is a input matrix which row represents the observation of the m-th task, $X = \left[X^1, X^2, \ldots, X^M\right]$ is the matrix ensemble concatenating input matrix on all tasks, $W^m \in R^D$ is the weight vector of the m-th task, $W \in R^{D\times M}$ is the weight matrix stacked by weight vectors of the all tasks. The $Y^m \in R^{N_m}$ is the output vector of the m-th task and we let $Y = \left[Y^1, \ldots, Y^M\right]$ be vector ensemble which tilts the output vectors of all tasks and the m-th column in Y is the response for task m, $\varepsilon_m \sim N(0, \sigma^2)$ represent the noise. Like the previous works [6, 7, 8, 9, 10], U stands for the similarity in task level while V retains the features other than shared structures.

According to our hypothesis, In task level $U = \left[U^1, \ldots, U^M\right]$ are divided into several clusters according to the whole similarity of U^m, and we apply multidimensional Gaussian mixture distribution $U^m \sim \sum_{k=1}^{K} \pi_k \mathcal{N}(b_k, \gamma_k^2)$ as its priori probability, the mixture has K components, U^m follows the normal distribution under each component:

$$P(U^m \mid b_k, \gamma_k^2) = \frac{1}{\sqrt{(2\pi)^D \mid \gamma_k^2 \mid}} \exp\left(-\frac{1}{2}(U^m - b_k)^T (\gamma_k^2)^{-1} (U^m - b_k)\right) \tag{2}$$

Where $b_k \in R^D$ is the mean of the k-cluster and $B = \left[b_1, \ldots, b_K\right]$ is the mean vector, $\gamma_k^2 \in R^{D\times D}$ is the covariance matrix in each mixture component, $\gamma^2 = \left[\gamma_1^2, \ldots, \gamma_K^2\right]$ is the covariance matrix array. $\pi = (\pi_1, \ldots, \pi_K)$ is the mixture component weights vector, meanwhile, we introduce an indicator variable matrix $C \in R^{M\times K}$, its row vector C_m indicates the class that the m-task belongs to and only one of it is taken as "1" and the rest is "0", the C_m follows multinomial distribution with parameter π:

$$C_m \sim M(\pi_1, \ldots, \pi_K) \Rightarrow p(c_{m,k} = 1) = \pi_k \tag{3}$$

Similar but different from U, we apply one-dimensional Gaussian mixture distribution $V_d^m \sim \sum_{g=1}^{G} \rho_g \mathcal{N}(q_g, \eta_g^2)$ to the every component V_d^m in V for describing the relationship between features. The G is the number of feature components and it is noted that the object of this distribution V_d^m is a scalar, so the mean q_g and variance η_g^2 of the g-th component are scalar, we define the mean vector $Q = (q_1, \ldots, q_G)$, the variance vector $\eta^2 = (\eta_1^2, \ldots, \eta_G^2)$, the weights vector $\rho = (\rho_1, \ldots, \rho_G)$, and the indicator variable $Z \in R^{(D\times M)\times G}$. The Z has $D \times M$ rows and G columns for it indicate all features, its element $(z_d^m)_g$ follows the multinomial distribution with parameter ρ_g like C: $(z_d^m)_g \sim M(\rho_1, \ldots, \rho_G)$ and $p((z_d^m) = 1) = \rho_g$. We draw probability graph of variables in Figure 4 (see Appendix 3).

Obviously, relationships in U^m or V_d^m is not independent because they are generated from a mixture of distributions, but we can write their joint distribution after introducing the indicator variable (the deduction is in Appendix 2) and acquire our generative probabilistic model as follows:

$$y_i^m = \sum_{d=1}^{D} x_i^m (U^m + V^m)^T + \varepsilon_m \Rightarrow P(Y \mid U, V, C, Z, \Theta) = \prod_{m=1}^{M} \prod_{i=1}^{N_M} \mathcal{N}(x_i^m (U^m + V^m)^T, \sigma^2)$$

$$P(U, C \mid \pi, B, \gamma^2) = \prod_{m=1}^{M} \prod_{k=1}^{K} \pi_k^{c_{m,k}} \mathcal{N}(U^m \mid b_k, \gamma_k^2)^{c_{m,k}} \tag{4}$$

$$p(V, Z \mid \rho, Q, \eta^2) = \prod_{d=1}^{D} \prod_{m=1}^{M} \prod_{g=1}^{g} \rho_g^{(Z_d^m)_g} \mathcal{N}(V \mid q_g, \eta_g^2)^{(Z_d^m)_g}$$

IOP Conf. Series: Journal of Physics: Conf. Series **1061** (2018) 012008 doi:10.1088/1742-6596/1061/1/012008

The bias w_0^m can be absorbed into x_i^m by adding an extra column in the matrix U^m, V^m and X. The $\Theta=\{\sigma^2,\pi,B,\gamma^2,\rho,Q,\eta^2\}$ is parameters to be estimated, because U, V, C, Z are also unknown parameters, we use MCEM to solve the parameters Θ in Appendix 4.

3. Experiments

In this section, we apply our algorithm to both synthetic and real datasets to compare with others state-of-the-art multi-task learning as follows: 1) Ridge regression method[12]; 2) convex feature learning[3] ($L_{2,1}$); 3) Clustered MTL[5] (CMTL); 4) Method for Trace Norm Minimization constraining weight matrix[6] (Low rank) ; 5) Dirty method with some sparse features[8] (Dirty); 6) robust multi-task learning[13] (rMTL);7) robust multi-task feature learning[7] (rMTFL); 8) convex multitask learning with flexible task clusters[9] (Flex-Clus); 9) Task-Feature Co-Clusters in Multi-Task Learning[10] (CoCMTL). The key evaluating criteria are the root mean squared error (RMSE), the normalization mean squared error (NMSE), and averaged mean squared error (AMSE).

The synthetic data: this section contains seven scenarios, in each scenario, in order to make sure it's a standard normalized form, the input x is generated from normal distribution $\mathcal{N}(0,1)$. We randomly generate 2000 examples and split it into 10 tasks, each task have 200 examples and each samples has 30 features, the number of training, validation, test samples is 30, 70 and 100, a single sample is represented as $y_i^m=(x_i^m)^T w^m+\mathcal{N}(0,10)$, where the last term is noise, according to the seven multi-tasking learning methods, we set seven types of W in virtual scenes to simulate the relationships:

Scene1: In this scene, all tasks are independent, $W^m \sim \mathcal{N}(\mu^m,I)$, $\mu^m \sim \mathcal{U}(0,5)$, $m=1,\text{L}\ 10$, $\mathcal{U}$ is the uniform distribution and $\mathcal{N}$ is Gaussian distribution, it is the same symbol in following scenes.

Scene2 (uniform feature structure): Tasks are related like $L_{2,1}$ [3], we guarantee the similarity of the inter-tasks by generating $W \sim \mathcal{N}(\mu,I)$ from the same distribution, where $\mu \sim \mathcal{U}(0,5)$.

Scene3 (task clustering): We make W obey several Gaussian probability clusters, and set three distributions to represent three clusters, i.e., $W^i \sim \mathcal{N}(\mu^i,I)$, $\mu^i \sim \mathcal{U}(0,5)$, $m=1,\text{L}\ 10$, $i=1,2,3$, each cluster contains 4, 3 and 3 tasks respectively, the difference between clusters are obviously.

Scene4(Flex-Clus): We simulate the special case of outlier features, for example, the recommendation system mentioned in flexible task clusters[9], we generate W from $\mathcal{N}(\mu,I)$, where $\mu \sim \mathcal{U}(0,5)$, then we replace the last two rows features $(W^i)_d$, $i=1,\text{L}\ ,10$ and $d=1,\text{L}\ ,30$ with $10+\mathcal{N}(0,10)$.

Scene5 (feature block): For simulating the scene in [10], we divide W into P and Q. P describes the similarity of the all tasks, which is generated from $\mathcal{N}(\mu,I)$, where $\mu \sim \mathcal{U}(0,1)$. On this foundation, 30 features are scattered into several rows: q_{1-5}, q_{6-15} and q_{16-30}. Then we sample 3, 4, 3 columns from $\mathcal{N}(\upsilon^i,I)$ separately to form blocks q^m , where $\upsilon^i \sim \mathcal{U}(0,10)$ and $m=1,2,3$. Q Is the block diagonalization of q^m , according to this assumption, the features in different blocks do not overlap and tasks are divided into different clusters by these feature blocks. Finally we got $W=P+Q$.

Scene6: The form of W and P is the same as scene5, but we want to simulate when there exist some relationships between these feature blocks. So we generate a column that elements have some relationship from a mixture distribution $q^i \sim \sum_{j=1}^{10}\mathcal{N}(\upsilon^j,1)$ where $i=1,\text{K}\ ,30$ and $\upsilon^j \sim \mathcal{U}(0,10)$, then we scatter and copy q^m as scene5 to form Q.

Scene7: The difference from above is: the task level P is arisen from a Gaussian mixture distribution $p^m \sim \sum_{i=1}^{4} \mathcal{N}(\mu^i, I)$ and $\mu^i \sim \mathcal{U}(0,1)$, there have a certain relationship with each other but their closeness is not inhomogeneous. On the basis of scene6, more general the situation is that features are not always form blocks and there are always some features left out, for simulating similar features in all feature spaces, we consider a more complex feature relationship, all features have a certain connection more or less, which means $(q^m)_d \sim \sum_{j=1}^{10} \mathcal{N}(\upsilon^j, I)$, $\upsilon^i \sim \mathcal{U}(0,10)$. We use ten components to describe it, then rearrange $(q^m)_d$ to D rows and M columns to get Q, finally we got $W = P + Q$.

There is no relation between tasks in scene1, so the experimental results (in Appendix 1) show Ridge achieve the best performance and most multitask learning methods has not enough worked, because the common structure is completely absent and it cause negative migration, but we can see the latter four methods focus on feature are next to Ridge, they learned some information due to only a few features can be migrated, so the performances are still better than methods focus on task level.

The result in scene2 shows that like $L_{2,1}$ or Low rank based on common structure have almost the better same performance because they are just different in the quantity of selected features, but the setting in scene2 have the same data structure and we have adopted the unit variance which means there are no probability relations between the features in every task.

In scene3 with the number of task clusters increasing, the prediction accuracy of the traditional methods basis on shared structure decline, CMTL performs well than the other overlapping models, because the P in those models based on the shared structure without considering the task clustering.

In the scene4, there appear two significant difference lines of features, which means there exist nonlinear relation between the feature vectors, because the significant change in the direction of features and most methods do not have the structural constraint, the prediction accuracy of uniform structured algorithm like $L_{2,1}$ norm, CMTL occur descend at different degrees. The model assumption of rMTL and rMTFL can conform to this scene because they focus on the isolate task or feature, besides the last three feature learning methods.

In these latter several scenes, accuracy of methods that only considering characteristics of the task descend, and the methods which considering characteristics of the features gradually became prominent. The rMTL and rMTFL does not deal with these forms of feature blocks. It is worth noting that although rMTL contain a low rank constraint on the matrix, but the other part of the overlapping is the shared structure constraint like $L_{2,1}$ norm, this part still produces negative migration, especially in the last three case it occur task clusters or more specific features, so when we compare rMTL with Low rank method, the latter is better. For scene5, Flex-clus perform well, but with the increase of the feature correlation in scene6 and scene7, CoCMTL and MTPFL give better results. The thermal map of the learned weight matrix is shown in figure 2 and result of the evaluating indicator is shown in table 1a and table 1b. We can see that MTPFL have advantage in feature learning, especially when these feature blocks are related, at the same time we can draw a conclusion that classification at the task level is necessary.

The school data: the dataset is consist of 15362 students from 139 schools, the exam score prediction of each schools can be seen as a task and each student was taken as a samples which contains 27 attributes, such as the year of admission, gender, ethnic and etc. Two attributes are in a representation of percentage number while the others are encoded into "0", "1", we also add an attribute to offset the bias W_0. Notice that the sample sizes in each task are different and the number of samples in some tasks is even less than the number of feature attributes, the result of single task regression is inaccurate, so the data set is suitable for MTL. We select 10%, 20% , 30% of the samples in each task as training set respectively, validation set is selected as 45%, and test sets are correspond to 45%, 35%, 25%.

The number of iterations is ten thousand times, after iteration, we can see from Figure 3 (In Appendix 1) although the training sets have different percentages, most irrelevant feature of U are induced into several clusters in task level, this is consistent with our hypothesis that task groups are more reasonable, and clusters are changing owing to the sample information varies with the capacity of different training sets. Meanwhile features of U became stable sparse because they are taken into consideration in V, some notable distinct lines show that these features are more strongly related to most tasks, which has an obvious auxiliary role for tasks with small sample capacity.

Matrix V mainly contains some unique features whose influence changing with tasks. It's very clear in Figure 3 that there are a lot of discrete features that belong to the same cluster. It is worth noting that, for example, the features in the eighth and ninth line are very dense, comparing the same positions of U, we can see that it is also significant in V, this is because according to our hypothesis, except for their influence in task level, they still have a part of their own characteristics, which is consistent with the property of the dataset because they are important features. The result shows that features between different tasks can be interrelated when the structure of V become gradually stable. And the comparison with other methods is in appendix 1 (Table 2).

Computer survey data: the dataset is from a real statistic data which also be used in many literature like [3], it describes an analysis of buying preferences for 20 brands of computers in the means of score "0-10", each brand contains 180 investigated customer samples that can be considered as separate tasks, attributes are 13 items configuration about computer like cpu, ram, cache, price, etc. Since these attributes are the same for each brand computer, that is, all samples (computer attributes) of the tasks (customer) are the same. Among the output set, we eliminate the samples which have too many invalid scoring items, eventually there are 172 samples left in the dataset. As we know, there are some correlation between these features, such as the quality of CPU directly affects the overall configuration, and better hardware are required to match it quickly, thus it affects the relationship on other features in the learning process. We verify it by experimental accuracy and MTPFL achieved a better performance, the results are shown in Appendix 1 (Table 3).

4. Conclusion

In this article, we propose a new generative model for multitask feature learning, the weight matrix is divided into two parts, and modeling the task and feature relationship simultaneously, according to the relationship between the task relation and feature modeling, and generative method can describe the relationship between these two parts. The selection of cluster number and the accuracy of sampling are still the research directions in the next work.

References

[1] Caruana R. Learning many related tasks at the same time with backpropagation// International Conference on Neural Information Processing Systems. MIT Press, 1994:657-664

[2] Evgeniou T, Pontil M. Regularized multi--task learning// Tenth ACM SIGKDD International Conference on Knowledge Discovery and Data Mining. ACM, 2004:109-117

[3] Argyriou, A.; Evgeniou, T.; and Pontil, M. 2008. Convex multi-task feature learning. Machine Learning 73(3):243–272

[4] Xue Y, Liao X, Carin L, et al. Multi-Task Learning for Classification with Dirichlet Process Priors . Journal of Machine Learning Research, 2007, 8(1):35-63

[5] Jacob L, Bach F, Vert J P. Clustered Multi-Task Learning: A Convex Formulation. Advances in Neural Information Processing Systems, 2008:745-752

[6] Ji S, Ye J. An accelerated gradient method for trace norm minimization// International Conference on Machine Learning, ICML 2009, Montreal, Quebec, Canada, June. DBLP, 2009:457-464

[7] Gong P, Ye J, Zhang C. Robust Multi-Task Feature Learning// Acm Sigkdd International Conference on Knowledge Discovery & Data Mining. KDD, 2012:895

[8] Jalali A, Sanghavi S, Ruan C, et al. A dirty model for multi-task learning[C]//Advances in neural information processing systems. 2010: 964-972

[9] Zhong W, Kwok J. Convex Multitask Learning with Flexible Task Clusters. 2012
[10] Xu L, Huang A, Chen E. Exploiting Task-Feature Co-Clusters in Multi-Task Learning. 2015
[11] Murugesan K, Carbonell J, Yang Y. Co-Clustering for Multitask Learning. 2017
[12] Arthur E. Hoerl, Robert W. Kennard. Ridge Regression: Biased Estimation for Nonorthogonal Problems. Technometrics, 1970, 12(1):55-67
[13] Chen J, Zhou J, Ye J. Integrating low-rank and group-sparse structures for robust multi-task learning// ACM SIGKDD International Conference on Knowledge Discovery and Data Mining. ACM, 2011:42-50

Appendix 1:

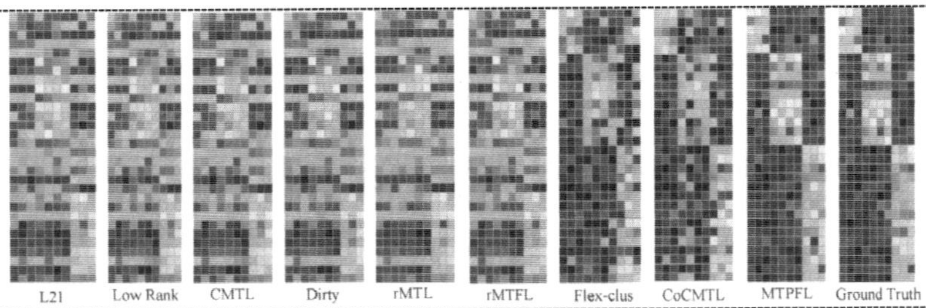

Figure 2. The thermal map of the learned weight matrix under these methods, structure learned by MTPFL is the closest to the Ground Truth.

Table 1a. Result on synthetic data. (The former five methods)

	ratio	Ridge	L21	Low rank	CMTL	dirty
	Syn1	2.5742[1]	2.9862[7]	2.9199[6]	3.1353[9]	3.2055[10]
	Syn2	2.3553[10]	2.1562[1]	2.1591[2]	2.1867[7]	2.2421[9]
	Syn3	3.3253[10]	3.1674[9]	2.8329[4]	2.5299[1]	3.1601[8]
RMSE	Syn4	2.9485[10]	2.6042[6]	2.5299[3]	2.7394[8]	2.8950[9]
	Syn5	2.7134[7]	2.7284[8]	2.3795[3]	2.6002[5]	2.9139[10]
	Syn6	2.5118[10]	2.2305[4]	2.0466[2]	2.3304[8]	2.3525[9]
	Syn7	3.3176[9]	3.2387[6]	2.9828[2]	3.2842[8]	3.2494[7]

Table 1b. Result on synthetic data. (The latter five methods)

	scene	rMTL	rMTFL	Flex-clus	CoCMTL	MTPFL
	Syn1	3.0742[8]	2.9101[5]	2.6039[3]	2.7929[4]	2.6016[2]
	Syn2	2.1936[8]	2.1851[6]	2.1698[4]	2.1742[5]	2.1606[3]
	Syn3	2.8290[3]	3.1203[7]	3.0425[6]	2.9745[5]	2.7551[2]
RMSE	Syn4	2.5337[4]	2.5440[5]	2.4213[2]	2.6795[7]	2.3837[1]
	Syn5	2.4311[4]	2.7327[9]	2.6345[7]	2.2751[2]	2.0738[1]
	Syn6	2.1547[3]	2.2482[6]	2.2490[7]	2.2405[5]	2.0112[1]
	Syn7	3.1225[4]	3.3606[10]	3.1453[5]	3.0967[3]	2.5124[1]

Table 2a. Result on school data. (The former five methods)

	ratio	Ridge	L21	Low rank	CMTL	dirty
	10%	11.4316[9]	11.2647[6]	11.2339[4]	11.3226[7]	11.4412[10]
RMSE	20%	11.0421[8]	11.0133[7]	10.7506[5]	10.7733[6]	11.0759[9]
	30%	10.6908[10]	10.5944[7]	10.5834[5]	10.5936[6]	10.6553[8]
	10%	1.0935[8]	0.9962[6]	0.9633[4]	0.9738[5]	1.1226[9]
NMSE	20%	1.0143[9]	0.8641[6]	0.8601[5]	0.9309[7]	0.9645[8]

	30%	0.9402[10]	0.8751[7]	0.8260[4]	0.8469[6]	0.8935[9]
	10%	0.2866[8]	0.2553[5]	0.2523[4]	0.2805[7]	0.2932[9]
AMSE	20%	0.2723[9]	0.2451[7]	0.2398[5]	0.2437[6]	0.2525[8]
	30%	0.2456[10]	0.2289[7]	0.2198[5]	0.2213[6]	0.2319[8]

Table 2b. Result on school data. (the latter five methods)

	ratio	rMTL	rMTFL	Flex-clus	CoCMTL	MTPFL
	10%	11.2515[5]	11.3735[8]	11.1491[2]	11.2221[3]	11.0653[1]
RMSE	20%	10.7361[4]	11.0771[10]	10.7037[3]	10.7013[2]	10.6369[1]
	30%	10.5755[4]	10.6731[9]	10.4154[2]	10.4971[3]	10.3615[1]
	10%	1.0694[7]	1.2982[10]	0.8983[3]	0.8357[1]	0.8532[2]
NMSE	20%	0.8393[4]	1.0389[10]	0.8121[3]	0.8092[2]	0.7961[1]
	30%	0.8413[5]	0.8901[8]	0.7970[3]	0.7849[2]	0.7703[1]
	10%	0.2606[6]	0.3394[10]	0.2282[1]	0.2389[2]	0.2447[3]
AMSE	20%	0.2363[4]	0.2813[10]	0.2172[2]	0.2297[3]	0.2123[1]
	30%	0.2161[4]	0.2325[9]	0.2057[2]	0.2080[3]	0.1948[1]

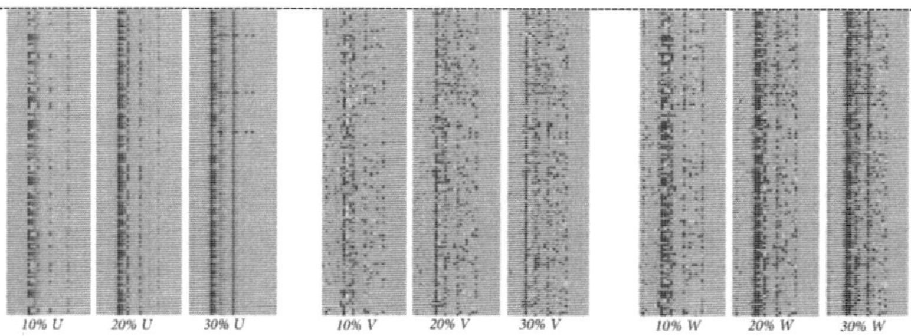

Figure 3. Comparison of *U, V* and *W*'s thermal map with different training sample percent in School data, the rows stand for tasks and columns stand for features. And close colors represent approximate relationship and they are maximum probability from the same probability cluster.

Table 3. Result on computer survey data.

Algorithm	NMSE	AMSE	RMSE
Ridge	2.4754	1.5162	2.1371
L2,1	2.4703	1.6324	2.1192
Low rank	2.1868	1.3651	2.0346
CMTL	2.2427	1.5567	2.0314
dirty	2.4341	1.6247	2.1257
rMTL	2.3679	1.4734	2.0425
rMTFL	2.3625	1.4029	2.0685
Flex-clus	1.9949	1.2459	2.0284
CoCMTL	1.8943	1.3392	2.0252
MTPFL	1.8683	1.2081	2.0142

Appendix 2:

The distribution of the task level vector U^m after introducing hidden variables C is:

$$p(U^m \mid \pi, B, \gamma^2) = \sum_{k=1}^{K} \pi_k \mathcal{N}(U^m \mid b_k, \gamma_k^2)$$

$$p(c_{m,k} = 1) = \pi_k \Rightarrow P(C_m) = \prod_{k=1}^{K} \pi_k^{c_{m,k}}$$

$$p(U^m \mid \pi, B, \gamma, c_{m,k} = 1) = \mathcal{N}(U^m \mid b_k, \gamma_k^2) \Rightarrow p(U^m \mid \pi, B, \gamma^2, C_m) = \prod_{k=1}^{K} \mathcal{N}(U^m \mid b_k, \gamma_k^2)^{c_{m,k}} \tag{1}$$

$$p(U, C \mid \pi, B, \gamma^2) = \prod_{m=1}^{M} p(C_m) p(U^m \mid \pi, B, \gamma^2, C_m) = \prod_{m=1}^{M} \prod_{k=1}^{K} \pi_k^{c_{m,k}} \mathcal{N}(U^m \mid b_k, \gamma_k^2)^{c_{m,k}}$$

The distribution of the feature level point V_d^m after introducing hidden variables Z is:

$$p(V_d^m \mid \rho, Q, \eta^2) = \sum_{g=1}^{G} \rho_g \mathcal{N}(V_d^m \mid q_g, \eta_g^2)$$

$$P((z_d^m)_g = 1) = \rho_g \Rightarrow P(Z_d^m) = \prod_{g=1}^{G} \rho_g^{(z_d^m)_g}$$

$$p(V_d^m \mid \rho, Q, \eta^2, (z_d^m)_g = 1) = \mathcal{N}(V_d^m \mid q_g, \eta_g^2) \Rightarrow p(V_d^m \mid \rho, Q, \eta^2, Z_d^m) = \prod_{g=1}^{G} \mathcal{N}(V_d^m \mid q_g, \eta_g^2)^{(z_d^m)_g} \tag{2}$$

$$p(V, Z \mid \rho, Q, \eta^2) = \prod_{d=1}^{D} \prod_{m=1}^{M} p(Z_d^m) p(V_d^m \mid \rho, Q, \eta^2, Z_d^m) = \prod_{d=1}^{D} \prod_{m=1}^{M} \prod_{g=1}^{G} \rho_g^{(z_d^m)_g} \mathcal{N}(V \mid q_g, \eta_g^2)^{(z_d^m)_g}$$

Appendix 3:

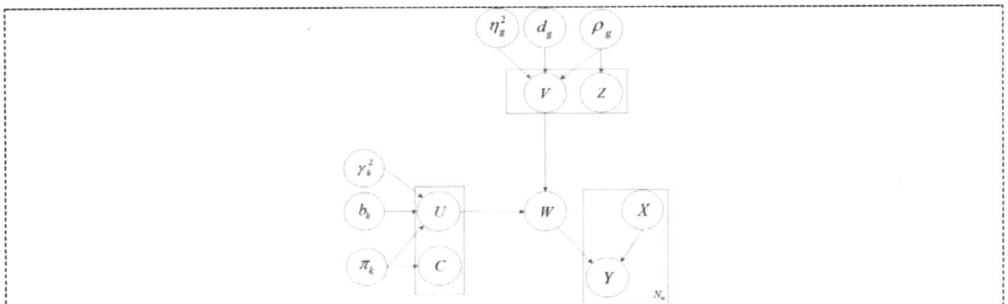

Figure 4. Graphical model for probabilistic weight matrix partition.

Appendix 4:

In this section we will adopt the Monte Carlo EM (MCEM) method to estimate the parameters $\Theta(\cdot)$ and derive parameters update equations. First we write the logarithm conditional posterior probability $\log p(Y \mid X, \Theta)$ after augmenting the data with U, V, C and Z as:

$$\log p(Y \mid X, \Theta) = \sum_C \sum_Z \int_u \int_v \log p(Y, U, V, C, Z \mid X, \Theta) d_u d_v \tag{3}$$

The initialization are needed before running the algorithm, the two mixture components proportion π and ρ are randomly allocated, γ^2 and η^2 are randomly assigned covariance ensembles, B and Q come from a uniformly distribution, and we randomly select rows in Z and C with "1", After the initial value of Θ has been determined, we sample from the distribution of $U^{(0)}$ and $V^{(0)}$, and in order to prevent occurring empty Gaussian components in the following iterations, we use the K-means method to recalculate $\pi^{(0)}, B^{(0)}, (\gamma^2)^{(0)}, \rho^{(0)}, Q^{(0)}, (\eta^2)^{(0)}$ as real initial values $\Theta^{(0)}$, then we determine $Z^{(0)}$ and $C^{(0)}$ according to result of the *K*-means. Then we formally enter the MCEM algorithm.

E-step: in the E-step, we calculate the approximate expectation of the log-likelihood:

$$Q(\Theta|\breve{\Theta}(t)) = E[\log p(Y,U,V,C,Z|X,\breve{\Theta}(t))|Y,X,\breve{\Theta}(t)] \approx \sum_{n=1}^{N}\log p(Y,\xi^{(n,t)},\varsigma^{(n,t)},\zeta^{(n,t)},\psi^{(n,t)}|X,\breve{\Theta}(t)) \tag{4}$$

On successive iterations, at *t*-th step of the iteration, to simulate *U, V, C, Z,* we implement repeatedly Gibbs sampling from conditional mathematical expectation $\xi^{(n,t)},\varsigma^{(n,t)},\zeta^{(n,t)},\psi^{(n,t)}$ of *U, V, C,* and *Z*:

$$\begin{aligned}
\xi^{(n,t)} &= E[U|Y,V,C,Z,X,\breve{\Theta}(t)] = Pr(U|V,C,Y^m,X^m,\breve{\Theta}(t))\\
\varsigma^{(n,t)} &= E[V|Y,U,C,Z,X,\breve{\Theta}(t)] = Pr(V|U,Z,Y^m,X^m,\breve{\Theta}(t))\\
\psi^{(n,t)} &= E[(z_d^m)_g|Y,\breve{\Theta}(t)] = Pr[(z_d^m)_g = 1|V_d^m,\breve{\Theta}(t)]\\
\zeta^{(n,t)} &= E[c_{m,k}|Y,\breve{\Theta}(t)] = Pr[c_{m,k} = 1|U^m,\breve{\Theta}(t)]
\end{aligned} \tag{5}$$

Note that the *W* is the sum of *U* and *V*, we construct generative models for the two components *U* and *V* respectively, the posterior distribution of *U* and *V* can be derived from the complete log-likelihood:

$$\begin{aligned}
L &= \log p(Y,U,V,C,Z|X,\Theta) = \log(Y|X,U,V,C,Z,\Theta) + \log p(U,C|V,Z,\Theta) + \log p(V,Z|\Theta)\\
&= \log(Y|X,U,V,\sigma^2) + \log p(U,C|\pi,b,\gamma^2) + \log p(V,Z|\rho,d,\eta^2)\\
&= L_{wuguan} - \frac{1}{2\sigma^2}\sum_{m=1}^{M}\sum_{i=1}^{N_m}\{y_i^m - \sum_{d=1}^{D}x_{i,d}^m(U_d^m + V_d^m)\}^2 + \sum_{m=1}^{M}\sum_{d=1}^{D}\sum_{g=1}^{G}(z_d^m)_g\{\log\rho_g - \frac{(V_d^m - q_g)^2}{2\eta_g^2}\}\\
&+ \sum_{m=1}^{M}\sum_{k=1}^{K}c_{m,k}\{\log\pi_k - \frac{1}{2}(U^m - b_k)^T(\gamma_k^2)^{-1}(U^m - b_k)\}\\
&= L_{wuguan} - \sum_{m=1}^{M}\{\frac{(Y^m)^TY^m - 2(U^m+V^m)^T(X^m)^TY^m}{2\sigma^2} + \frac{(U^m+V^m)^T(X^m)^TX^m(U^m+V^m)}{2\sigma^2}\\
&+ \frac{(U^m)^TU^m - 2(U^m)^TC_mB}{2C_m\gamma^2} + \frac{(V^m)^TV^m - 2(V^m)^TZ_d^mQ}{2Z_d^m\eta^2}\}\\
&= K(\cdot) - \sum_{m=1}^{M}\{\frac{(U^m)^T(X^m)^TX^mU^m}{2\sigma^2} + \frac{(U^m)^T(X^m)^TX^mV^m}{\sigma^2} - \frac{(U^m)^T(X^m)^TY^m}{\sigma^2} + \frac{(U^m)^TU^m - 2(U^m)^TC_mB}{2C_m\gamma^2}\}\\
&= F(\cdot) - \sum_{m=1}^{M}\{\frac{(V^m)^T(X^m)^TX^mV^m}{2\sigma^2} + \frac{(V^m)^T(X^m)^TX^mU^m - 2(V^m)^T(X^m)^TY^m}{\sigma^2} + \frac{(V^m)^TV^m - 2(V^m)^TZ_d^mQ}{2Z_d^m\eta^2}\}
\end{aligned} \tag{6}$$

The $K(\cdot)$ and $F(\cdot)$ contain mathematical variables that are independent from *V* and *U* respectively. The process solving posterior distribution about *U* need to separate the related parameters from likelihood functions (6) as follows:

$$\begin{aligned}
L &= K(\cdot) - \sum_{m=1}^{M}\{\frac{(U^m)^T(X^m)^TX^mU^m + 2(U^m)^T(X^m)^TX^mV^m}{2\sigma^2} - \frac{(U^m)^T(X^m)^TY^m}{\sigma^2} + \frac{(U^m)^TU^m - 2(U^m)^TC_mB}{2C_m\gamma^2}\}\\
&= K(\cdot) - \frac{1}{2\sigma^2}\{(U^m)^T[(X^m)^TX^m + \frac{\sigma^2}{C_m\gamma^2}]U^m - 2(U^m)^T[\frac{\sigma^2C_mB}{C_m\gamma^2} + (X^m)^TY^m - (X^m)^TX^mV^m]\}\\
&= K(\cdot) - \frac{1}{2\sigma^2[(X^m)^TX^m + \frac{\sigma^2}{C_m\gamma^2}]^{-1}}\{(U^m)^TU^m - 2(U^m)^T\frac{[\frac{\sigma^2C_mB}{C_m\gamma^2} + (X^m)^TY^m - (X^m)^TX^mV^m]}{[(X^m)^TX^m + \frac{\sigma^2}{C_m\gamma^2}]}\}\\
&= K(\cdot) - \frac{1}{2\sigma^2[(X^m)^TX^m + \frac{\sigma^2}{C_m\gamma^2}]^{-1}} \times \left\{(U^m)^T - \frac{[\frac{\sigma^2C_mB}{C_m\gamma^2} + (X^m)^TY^m - (X^m)^TX^mV^m]}{[(X^m)^TX^m + \frac{\sigma^2}{C_m\gamma^2}]}\right\}^2
\end{aligned} \tag{7}$$

IOP Conf. Series: Journal of Physics: Conf. Series **1061** (2018) 012008 doi:10.1088/1742-6596/1061/1/012008

Note that U defines the weight matrices, the sampling process of U is vector by vectors, according to our hypothesis, the posterior distribution $p(U^m \mid C,V,y^m,x^m,\Theta) \sim \mathcal{N}(\mu_U,\Sigma_U)$ still follows Gaussian distribution, from (7) the mean and variance of U^m are:

$$\begin{aligned}\mu_U &= [(X^m)^T X^m + \sigma^2 C_m (\gamma^2)^{-1} I_D)]^{-1} \times \{\sigma^2 C_m B [C_m(\gamma^2)]^{-1} + (X^m)^T Y^m - (X^m)^T X^m V^m\} \\ \Sigma_U &= \sigma^2 [(X^m)^T X^m + \sigma^2 C_m (\gamma^2)^{-1} I_D]^{-1}\end{aligned} \tag{8}$$

The solving process about posterior distribution $p(V_d^m \mid Z,Y,X,\Theta) \sim \mathcal{N}(\mu_V,\Sigma_V)$ is similar as U, but V is relation to elements in the matrix, we need change structural of X,Y,U,V to $X_{blk},Y_{blk},U_{blk},V_{blk}$:

$$X_{blk} = \begin{Bmatrix} X^1 & & & \\ & X^2 & & \\ & & 0 & \\ & & & X^M \end{Bmatrix} \quad Y_{blk} = \begin{Bmatrix} Y^1 & & & \\ & Y^2 & & \\ & & 0 & \\ & & & Y^M \end{Bmatrix} \quad U_{vec} = ((U^1)^T,(U^2)^T,\mathrm{K}\,(U^M)^T) \quad V_{vec} = ((V^1)^T,(V^2)^T,\mathrm{K}\,(V^M)^T) \tag{9}$$

And the mean and variance of V_d^m is:

$$\begin{aligned}\mu_V &= ((X_{blk})^T X_{blk} + \frac{\sigma^2}{Z_d^m \eta^2} I_{D\times M})^{-1} \times \{\frac{\sigma^2 Z_d^m Q}{Z_d^m \eta^2} + (X_{blk})^T Y_{blk} - (X_{blk})^T X_{blk} U_{vec}\} \\ \Sigma_V &= \sigma^2 ((X_{blk})^T X_{blk} + \frac{\sigma^2}{Z_d^m \eta^2} I_{D\times M})^{-1}\end{aligned} \tag{10}$$

C and Z are the probability that the data is generated by every Gaussian component. So the scales of ψ,ζ are the same as C,Z , the probability that U^m is generated by the k-th Gaussian components is:

$$\psi_k^m = E[c_{m,k} \mid Y,\overset{\smile}{\Theta}(t)] = \Pr[c_{m,k} = 1 \mid U^m,\overset{\smile}{\Theta}(t)] = \frac{\pi_k \mathcal{N}(U^m \mid b_k,\gamma_k^2)}{\sum_{k=1}^K \pi_k \mathcal{N}(U^m \mid b_k,\gamma_k^2)} \tag{11}$$

Similarly, the probability that $Vvec$ is generated by the g-th Gaussian components is:

$$(\zeta_d^m)_g = E[(z_d^m)_g \mid Y,\overset{\smile}{\Theta}(t)] = \Pr[(z_d^m)_g = 1 \mid V_d^m,\overset{\smile}{\Theta}(t)] = \frac{\rho_g \mathcal{N}(V_d^m \mid q_g,\eta_g^2)}{\sum_{g=1}^G \rho_g \mathcal{N}(V_d^m \mid q_g,\eta_g^2)} \tag{12}$$

Up to now, we have derived the conditional posterior distribution of four hidden auxiliary matrices U, V, C and Z, and then we can implement repeatedly Gibbs sampling.

M-step: on the basis of obtained the updated values of unobserved matrices in E-step, In the M-step, to maximize the $Q(\Theta \mid \overset{\smile}{\Theta}(t))$ and get updating parameters $\overset{\smile}{\Theta}(t+1)$, we take the derivatives of the likelihood function and set equal to zero, and recalculate the weight, variance and mean of each clusters. Meanwhile we define the probability of number of components in the clusters:

$$N_k = \sum_{m=1}^M \psi_{m,k} \quad N_g = \sum_{m=1}^M \sum_{d=1}^D (\zeta_d^m)_g \tag{13}$$

Then the updating formulas of the parameters $\Theta = \{\sigma^2,\pi,B,\gamma^2,\rho,Q,\eta^2\}$ are:

$$\begin{aligned}&\sigma^2 = \sum_{m=1}^M \sum_{i=1}^{N_m} (Y^m - X^m(U^m + V^m))^T \times (Y^m - X^m(U^m + V^m)) \\ &\pi_k = \frac{\sum_{m=1}^M \psi_k^m}{M} \qquad b_k = \frac{1}{N_k}\sum_{m=1}^M \psi_k^m U^m \qquad \gamma_k^2 = \frac{1}{N_k}\sum_{m=1}^M \psi_k^m (U^m - b_k)(U^m - b_k)^T \\ &\rho_g = \frac{\sum_{m=1}^M \sum_{d=1}^D \zeta_{d,g}^m}{D\times M} \qquad q_g = \frac{1}{N_g}\sum_{m=1}^M \sum_{d=1}^D \zeta_{d,g}^m V_d^m \qquad (\eta_g^2)^{(i)} = \frac{1}{N_g}\sum_{m=1}^M \sum_{d=1}^D \zeta_{d,g}^m (V_d^m - q_g)(V_d^m - q_g)^T\end{aligned} \tag{14}$$

Automated Toll Collection System based on Vehicle Type Classification using Sparse-Filtered Convolutional Neural Networks with Layer-Skipping Strategy (SF-CNNLS)

Suryanti Awang[1] and Nik Mohamad Aizuddin Nik Azmi[1]

[1]Soft Computing & Intelligent System Research Group (SPINT), Faculty of Computer Systems & Software Engineering, Universiti Malaysia Pahang, Lebuhraya Tun Razak, 26300, Pahang, Malaysia

*suryanti@ump.edu.my

Abstract. Automated Toll Collection System (ATCS) is one of the technologies to fulfill the Intelligent Transportation System's (ITS) aim in providing an efficient road and transportation infrastructure at the expressway. This paper is aimed to provide an accurate and efficient ATCS based on a vehicle type classification method rather than the current implementation of toll collection that rely on sensor-based and human observation. To fulfill the aim, we proposed to implement SF-CNNLS framework to extract vehicle's features and classify it into class 1 (passenger vehicle), class 2 (lorry) and class 4 (taxi). This ATCS is aimed to enhance the efficiency of the toll collection in Malaysia. The biggest challenge in this research is how to discriminate features of class 4 as a different class of class 1 since both classes have almost identical features. However, with our proposed method, we able to classify the vehicle with the average accuracy of 90.83 %. We tested our method using a frontal view of a vehicle from the self-obtained database (SPINT) taken using mounted-camera at the toll booth and compare the classification performance with a benchmark database named BIT.

1. Introduction

Intelligent Transportation System (ITS) corresponds to the technology applied in road and transportation infrastructure to improve safety and efficiency in related applications, for instance, expressway toll system, traffic census, traffic surveillance, etc. This paper focuses on the expressway toll system especially in Malaysia, conventional methods, for instance, manually observed by human, sensor-based and electronic tools based is implemented. Basically, the expressway toll collection system will charge the vehicle based on the classes; class 1 for passenger vehicle (sedan car, MPV, SUV, etc), class 2 is for lorry, class 3 is for truck, class 4 is for taxi and class 5 is for bus. One of the issues with the current implementation of the toll collection system is human error may occur during the observation. Furthermore, as for the sensor-based and electronic tools based implementation, they are high in maintenance and need some configuration tools as in some of the existing works [1,2]. Therefore, we proposed to overcome the issues by using ATCS that is camera-based (computer vision and combined with the advantages of machine learning). The advantages of this approach is low in maintenance and easy to be configured since only camera and a system are needed to classify the vehicle.

There are a few works in ATCS that implemented computer vision to classify the vehicle class. For example, [3] detects the vehicle for toll collection using computer vision with embedded Linux. The vehicle is classified into light and heavy vehicle based on the vehicle image captured using a mounted-camera at the toll booth. Different from work done by [4], the authors recognized license plate of the vehicle that use an expressway at the toll booth in classifying the class of vehicle by using a segmentation approach and template matching in the image processing.

Since there is very limited of the existing ATCS work that implement the computer vision and machine learning, we considered the works in vehicle counting system (VCS) that are much related to this research. For example, a surveillance camera and traffic camera from top view are used to track and count vehicles in works done by [5,6,7,8]. The vehicles are counted as general (vehicle or non-vehicle) without any specific classification. These approaches are able to count the vehicles, however, unable to give an accurate outcome as expected in the traffic census. [9] also counts the vehicle without classifying its class using a motion estimation with Taylor series approximation. Other than that, there is Principle Component Pursuit (PCP) is used to extract features of vehicles from a satellite image, and vehicles are counted based on that features and does not consider the class of vehicle [10].

There is a VCS that is based on vehicle types proposed by [11]. They count the vehicles as small car, van, and motorcycle using Gaussian Mixture Model (GMM) to detect the vehicle. The drawback of this approach is the feature that extracted to be used in classifying the vehicle is too limited which is only rely on the size of blob. VCS also can be developed by using a combination of techniques in computer vision and machine learning. [12] proposed to implement this combination in VCS for traffic control analysis. They applied Fast Region-based Convolutional Neural Network (FR-CNN) to detect the moving vehicles. However, this approach only detects and count the moving vehicles without able to do the counting based on vehicle types.

Based on the existing works, we found that the ATCS can be improved with the implementation of intelligent techniques in machine learning to classify the vehicle type. Therefore, we proposed to implement SF-CNNLS based on deep learning method. This framework is able to learn vehicle features in detail and the ATCS is classified based on the classes that defined by the expressway concession. However, we notice that the main challenge in this approach is how to differentiate class 1 and class 4. This is because the features of both class is almost similar especially in a grayscale image. Thus, our aim is to overcome that challenge. The explanation on the methodology of ATCS with SF-CNNLS is in the next section, followed by experiment and result section and end with conclusion section.

2. Methodology

There are two stages of methodology in this paper; ATCS and vehicle type classification. We will explain the ATCS process flow first, and later is the explanation on vehicle type classification based on SF-CNNLS framework.

2.1. Automated Toll Collection System (ATCS)

We captured the vehicle image from a mounted camera video to be classified based on the vehicle class and toll rate. For instance, of toll rate with a flat rate for each class taken from the specific toll plaza, whereby, the expressway concession divided the passenger vehicle as class 1, class 2 for lorry, class 3 for truck or heavy vehicle, class 4 for taxi and class 5.

To detect the vehicle in the video, we use region based in detecting the region of interest for moving vehicle. The region is allocated in a specific in the video to capture a moving vehicle one by one. Figure 1 shows the process flow of ATCS. For the first phase, we consider to classify the vehicle from class 1, class 2 and class 4. From the video, when a vehicle is detected, vehicle type classification will classify it using SF-CNNLS. The ATCS will display the toll rate according to the class.

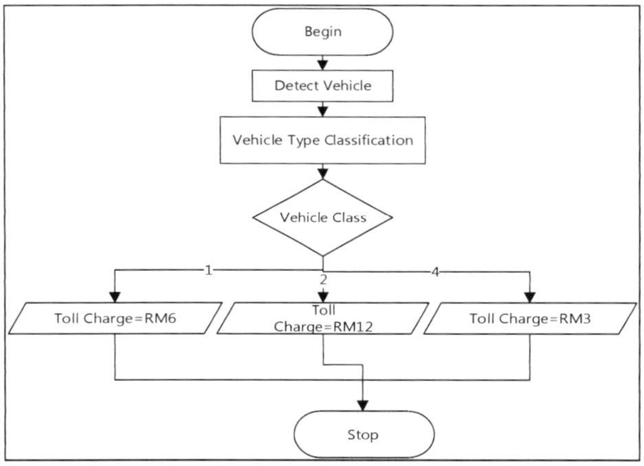

Figure 1. ATCS process flow.

2.2. Vehicle Type Classification –SF-CNNLS

This section is about how SF-CNNLS is implemented in classifying the vehicle class. There are two processes, namely, image acquisition to acquire vehicle images and pre-processing to prepare the images prior the implementation of SF-CNNLS. In that processes, the training and testing images that contain frontal-view of the vehicle are recorded using a surveillance camera. The pre-processing technique that has been used is a combination of existing works done by [11,12]. There are five pre-processing methods used, which are converting from RGB color space to grayscale, histogram equalization (HE), resizing with maintained aspect ratio, normalizing to zero mean and unit variance, and local contrast normalization (LCN) consequently. The SF-CNNLS is used as a feature extraction technique to extract vehicle features to be classified in the classification process. The SF-CNNLS is a sparse-filtered of the convolutional neural networks with layer-skipping strategy that has capability to extract both local and global features of a vehicle. Figure 2 shows the overview framework of SF-CNNLS implementation.

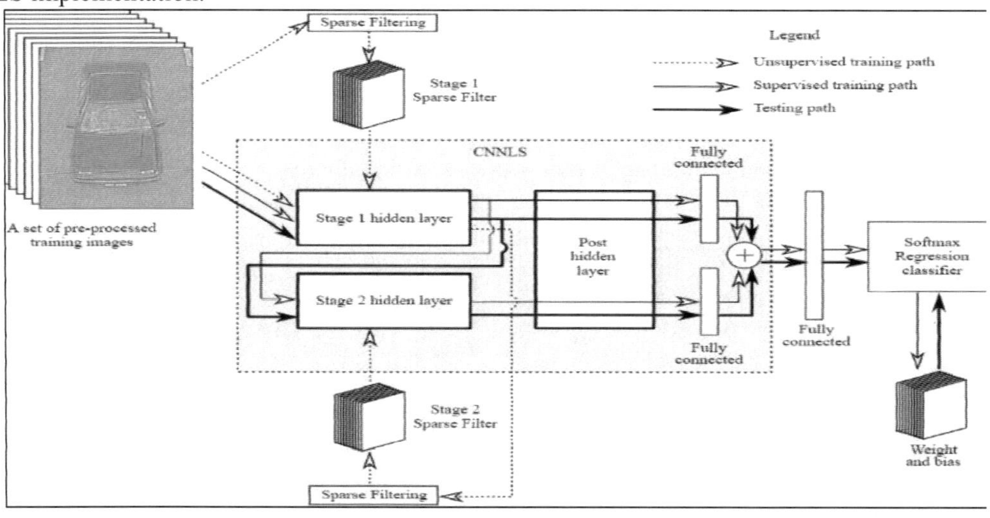

Figure 2. SF-CNNLS Overview Framework

There are 2 major phases in the SF-CNNLS, namely, training and testing phase. In the CNNLS process, there are two stages of hidden layer, one post hidden layer and fully connected that will implement the feature extraction of the pre-processed images. Note that there are two types of training phase, namely, unsupervised and supervised. The purpose of the unsupervised training is to generate two stages of optimized Sparse Filters, whereas the supervised training is to generate parameter weights and biases that will be used by the classifier in the classification process.

During the unsupervised training, a set of pre-processed images are firstly delivered into the Sparse Filtering function to generate a set of optimized stage 1 sparse filters. Later, the set of pre-processed input images are delivered into CNNLS stage 1 hidden layer and convolved with the optimized stage 1 sparse filters. The output from the CNNLS stage 1 hidden layer is used as an input for the Sparse Filtering function to generate a set of optimized stage 2 sparse filters. Thus, the outcome of the unsupervised training is a set of optimized stage 1 sparse filters and a set of optimized stage 2 sparse filters.

For the supervised training, the set of pre-processed images are delivered into CNNLS stage 1 hidden layer and will be convolved with the stage 1 sparse filters. The extracted features from this stage will be an input to the stage 2 hidden layer and the post hidden layer. In the stage 2 hidden layer, the input is convolved with the stage 2 sparse filters and the same layers are deployed to obtain the extracted features. This extracted features will be an input to the post hidden layer. The output from both stages which are the extracted features will be concatenated at the fully connected process into a single vector. This single vector feature is used to train the Softmax Regression classifier. Note that the testing set in testing phase has similar process with the supervised training except at the classifier where the trained weights and biases are used to calculate classification probabilistic in obtaining a classification result.

2.2.1. Feature Extraction: SF-CNNLS and Classification using Softmax Regression. There are five components called layers in each of the hidden layer. The components are convolutional layer, absolute value rectification (AVR) layer, local contrast normalization (LCN) layer, average pooling layer and subsampling layer without zero padding [13]. The output from each layer is the extracted features that will be an input features to another layer accordingly.

In the convolutional layer, the pre-processed image will be convolved with the optimized Sparse Filters. The purpose of convoluting the image is to extract the image features. From here, the convolved images are produced and a sigmoid activation function is applied on each convolved image. The next component is AVR layer. AVR is inspired from biological system that human eyes do not perceived images in negative values. This layer applies absolute value operation on the extracted features from the previous layer and the output will have absolute value elements. In the next component, the LCN that applied during CNNLS hidden layers is different compared to the LCN applied during the pre-processing process. Both LCNs has similar subtractive and divisive operations except that the input is a set of extracted features from the convolutional layer and different maximum value. The purpose of applying an average pooling after the LCN layer on the extracted features is to ensure that the extracted features are robust to geometric distortion. Thus, the features become less sensitive to a variation in angle and size of a vehicle. The last layer in the hidden layer is the subsampling layer without zero padding. The procedure in this layer is similar to the procedure of resize with the maintained aspect ratio in the pre-processing process.

The components that involved in the post hidden layer are AVR and subsampling layer with zero padding. The same process of AVR in the hidden layer is applied in this layer. The subsampling layer with zero padding is applied on the extracted features to ensure every feature has the same size and has the aspect ratio of 1:1 before being delivered into Softmax Regression classifier. The last process of feature extraction in CNNLS is the fully connected. A fully connected vector is a vector that its elements are in 1-dimensional.

Classifying vehicles is performed by executing a Softmax Regression hypothesis function. The hypothesis is calculated for each vehicle classes to find the amount of probabilities that the extracted features belong to which class. There are two phases of the Softmax Regression implementation; supervised training and testing. The Softmax Regression for the supervised training is trained to produce optimum weights and biases that will be used in the testing phase where it will be loaded and the hypothesis is calculated instead of minimizing the negative log-likelihood. The testing is performed on each vehicle class dataset.

3. Result and Discussion

We test our ATCS with the video captured from a mounted-camera with a frontal view of vehicle in a single lane and during daylight. The distance between the detected vehicle image in the video and the toll booth is 10 meter. Details of database gathering is explained in [13]. The processing time is recorded in classifying the vehicles to observe the efficiency of ATCS. This database is known as SPINT vehicle database. Note that the total samples for each vehicle class in the training dataset is 100 images, thus overall total is 300 images. **Table 1.** Training and Testing Samples Distribution for Performance Comparison

During our video recording for this experiment, we able to capture 65 vehicles for class 1, 19 vehicles for class 2 and 20 vehicles for class 4 for the testing purpose. Based on that, we produce a confusion matrix to the performance of classification for all vehicle type in the ATCS. We also did a performance comparison in terms of the vehicle type classification using a benchmark data from BIT vehicle database [12]. Note that for SPINT dataset, we used the same testing setup as the ATCS experiment. We summarize both of the experimental setup in table 1 below. The table shows the distribution of dataset in the training and testing for each vehicle class for each database. It shown that the taxi class which is class 4 has fewer samples in the BIT dataset, whereas, the truck which is class 2 has fewer samples in the SPINT dataset. For condition of the vehicle images, BIT provides captured images from frontal angle of surveillance camera during daylight similar to SPINT, and with slight illumination and shadow.

Table 1. Training and Testing Samples Distribution for Performance Comparison

Dataset		Class		
		1	2	4
BIT	Training	100	100	100
	Testing	67	21	17
SPINT	Training	100	100	100
	Testing	65	19	20

Table 2 shows the confusion matrix result of the average correctly classified percentages and misclassified percentages for both experiments. Looking at the confusion matrix, for ACTS result based on SPINT dataset, class 2 has the highest correctly classified percentage with the average of 99%, followed by class 2 with 98.5% and class 4 with 75%. Thus, the average of correctly classified is 90.83%. Class 2 has outperformed other classes where it has been misclassified as class 1 only with 1.05% and 0% for class 4. For the processing time, we found that the average time to classify one vehicle is 5.8 second. It is reasonably efficient in terms of the classification and display the toll rate process when we consider the distance of camera and toll booth.

Table 2. Confusion Matrix of Vehicle Type Classification

		Predicted					
		Class 1		Class 2		Class 4	
		BIT	SPINT	BIT	SPINT	BIT	SPINT
Actual	Class 1	92.5%	98.5%	4 %	6%	3.5%	0%
	Class 2	4%	1.05%	90%	99%	6%	0%
	Class 4	11%	19%	0%	0%	89%	75%

Furthermore, using BIT dataset, our approach is able to correctly classify the vehicle class respectively. The most important is class 4 is able to be classified as different class from class 1 in both datasets with 89% and 75%, respectively. However, 11% and 19% of the vehicle in class 4 is misclassified as class 1 in which the highest misclassified rate compared to other classes. Overall, SPINT dataset produced a better result compared to BIT due to the illumination in SPINT dataset is less than BIT dataset.

4. Conclusion
We proposed the implementation of Automated Toll Collection System (ATCS) based on vehicle type classification using deep learning method which is SF-CNNLS to classify the vehicle class in determine the toll rate. The aim is to provide an accurate classification result for ATCS. It also to overcome the existing issues of the toll collection system as mentioned earlier. Based on our approach, we able to automate the classification and toll collection process by classifying the vehicle automatically into class 1, class 2 and class 4. We proved that class 4 can be classified as different class from class 1 which is none of the existing works have done this. As mentioned in section 3, class 1 being misclassified as class 4 and vice versa is significantly high, this is due to lack of unique features to differentiate the class. Thus, the approach need to be improved especially in extracting the features to significantly differentiate the class. We also compare the vehicle type classification performance with the benchmark database. Based on that result, self-obtained database (SPINT) produce a better result compared to the benchmark database (BIT). In future, the processing time need to be increased in order to enhance the efficiency of the ATCS due to some camera is installed in shorter distance to the toll booth, for example, 5 meter. Thus, with our average processing time for one vehicle, this approach is required for an improvement especially in the feature extraction.

Reference
[1] Nagothu S K 2016 Automated toll collection system using GPS and GPRS. *International Conference on Communication and Signal Processing (ICCSP)* Melmaruvathur pp. 0651-0653.

[2] Tan J Y, Ker P J, Mani D and Arumugam P 2016 Development of a GPS-based highway toll collection system *6th IEEE International Conference on Control System, Computing and Engineering (ICCSCE)* Batu Ferringhi pp. 125-128.

[3] Suryatali A and Dharmadhikari V B 2015 Computer vision based vehicle detection for toll collection system using embedded Linux *International Conference on Circuits, Power and Computing Technologies* Nagercoil, pp. 1-7.

[4] Soomro S R, Javed M A and Memon F A 2012 Vehicle Number Recognition system for automatic toll tax collection *International Conference of Robotics and Artificial Intelligence* Rawalpindi pp. 125-129.

[5] Subaweh M B and Wibowo E P 2016 Implementation of Pixel Based Adaptive Segmenter method for tracking and counting vehicles in visual surveillance *International Conference on Informatics and Computing (ICIC)* Mataram, Indonesia, pp. 1-5.

[6] Kamkar S and Safabakhsh R 2016 Vehicle detection, counting and classification in various conditions IET Intelligent Transport Systems, vol. 10, no. 6, pp. 406-413.

[7] Bautista, C. M., Dy, C. A., Manalac, M. I., Orbe, R. A., and Cordel, M. 2016. Convolutional neural network for vehicle detection in low resolution traffic videos. In 2016 IEEE Region 10 Symposium (TENSYMP), pages 277–281.

[8] Valiere, P., Khoudour, L., Crouzil, A., and Cong, D. N. T. 2015. Robust vehicle counting with severe shadows and occlusions. In International Conference on Image Processing Theory, Tools and Applications (IPTA), Orleans, pp. 99-104.

[9] Prommool, P., Auephanwiriyakul, S., and Theera-Umpon, N. 2016. Vision-based automatic vehicle counting system using motion estimation with Taylor series approximation. In 6th IEEE International Conference on Control System, Computing and Engineering (ICCSCE), Penang, Malaysia, pp. 485-489.

[10] Quesada, J., and Rodriguez, P. 2016. Automatic vehicle counting method based on principal compo-nent pursuit background modeling. In IEEE International Conference on Image Processing (ICIP), Phoenix, AZ, pp. 3822-3826.

[11] Zhang, Z., Liu, K., Gao, F., Li, X., and Wang, G. 2016. Vision-based vehicle detecting and counting for traffic flow analysis. In International Joint Conference on Neural Networks (IJCNN), Vancouver, BC, pp. 2267-2273.

[12] Dong, Z., Pei, M., He, Y., Liu, T., Dong, Y., and Jia, Y. 2014. Vehicle type classification using unsupervised convolutional neural network. In 22nd IEEEE International Conference on Pattern Recognition(ICPR), pages172–177.

[13] Awang S., and Azmi N.M.A.N. 2018. Vehicle Counting System Based on Vehicle Type Classification Using Deep Learning Method. In: Kim K., Kim H., Baek N. (eds) IT Convergence and Security 2017. ICITS 2017.Lecture Notes in Electrical Engineering, vol 449. Springer, Singapore.

IOP Conf. Series: Journal of Physics: Conf. Series **1061** (2018) 012010 doi:10.1088/1742-6596/1061/1/012010

The Application of Wifi Technology in Smart Home

Jie Ding[1], Tian-ran Li[1] and Xue-Li Chen[1]

[1]Nanjing Normal University,Nanjing ,China

*E-mail: 540140028@qq.com

Abstract. With the continuous maturity of WiFi technology and the growing development of smart home, the use of WiFi technology to control home systems on smartphones will become an important part of people's lives in the future. More and more enterprises have joined the industries of developing smart home devices. More and more enterprises have joined the ranks of developing smart home devices. Firstly, this article introduces the concept of WiFi technology and the concept of smart home, and expounds its development trend. Secondly, it introduces the overall structure design of the system. Finally, it describes the hardware and software design of the system.

1. Introduction

With the rapid development of science and technology, the world has stepped into the era of informationization. Due to the continuous improvement of technology, people are more comfortable with home conditions and are more convenient and safer. As a result, smart buildings and homes are spawned. Smart home is through a home network of bus home devices connected to the information related to the implementation of network monitoring and control in order to manage centralized or off-site. At present, the development trend of smart home is from centralized control to distributed control. There are four mainstream wireless technologies currently used in smart home: Bluetooth, WiFi, Z-Wave, Zig-bee. The most competitive technology is WiFi technology and Zig-bee [1]. Therefore, this article will introduce the smart home control system based on WiFi.

2. WiFi technology principle

The full name of WiFi is Wireless Fidelity, it is 802.11b wireless network specification and a short-range wireless transmission technology that aims to improve the interoperability of wireless network products based on the IEEE802.11 standard. WiFi typically uses 2.4G UHF or 5G SHF ISM RF bands. Wireless LAN connections are usually protected by password, but they can also be open, passwordless, and network-wide devices can be connected.

The principle of the smart home control system based on WiFi is that a number of wireless smart nodes form a home area network, each smart sub-node includes a WiFi receiving module, and the home gateway has a wireless WiFi transmitting module. Receive and transmit module communicate through WiFi.

The advantages of WiFi are as follows:

- It covers a wide area and has a long effective distance. The communication distance in the open area is up to 305m, and the communication distance in the closed area is 76-122m. It can be applied not only to the local rooms but also to the entire buildings.

- Transmission speed is very fast, up to 54Mbps. In addition to the low cost, it has low power consumption, in line with the modern "green low-carbon" appeal. According to the research, the WiFi signal belongs to a kind of microwave and has little impact on the human body. The impact of the radiation generated by Wi-Fi on the human body is negligible within a safe range.

However, since WiFi uses RF technology, air is used as a medium to transmit and receive data, signals that transmitted through radio waves are easily disturbed by external environments, and the security is not high, so they are not suitable for smart home control systems that require high security.

3. Smart home

3.1. Concept and development process

The concept of smart home originated in the United States in the early 1980s, calling it Smart Home. It has undergone four generations of development: the first generation completes home networking through coaxial lines and two cores to achieve lighting, curtains and a small amount of security control; The second generation is through bus and IP technology networking, to be able to complete the video intercom and security business; The third generation is a centralized intelligent control system, by the central control computer to complete security, measurement and other functions; The fourth generation that is based on the Internet of things technology can be based on user needs to achieve personalized features [2].

Smart home should be integrated into four parts: building automation systems, communications systems, room automation systems, computer-aided equipment management system [3]. Smart home control system has intelligent, information-oriented functions, control functions, communication functions, monitoring functions, timing functions and abnormal protection, leakage protection [4]. China introduced the concept of smart home from the late 1980s. China started to promote the construction of intelligent residential quarters through the concept and technology of intelligent buildings. Adopting the intelligent residential district system mode of "smart residential as the basic unit, intelligent building's management center as the central node, and Hybrid Fiber Coax(HFC) network as the data transmission medium", it has the functions of Safety Automation (SA), Communication Network (CN) and Management Automation (MA) and other functions that Satisfy residential needs.[5]

3.2. Development Opportunities

Thanks to socio-economic progress, government support and technological development and business competition, smart home has been able to develop rapidly in recent years.

3.2.1. *Social and economic progress.* Under the influence of the reform and opening up and economic globalization, the economy of our country has shifted from a high-speed growth stage to a high-quality development stage. As the economic level continues to rise, people's demands for living quality are getting higher and higher. Nowadays, the younger generation is more receptive to advanced technologies, which provides a solid foundation for the development of smart home.

3.2.2. Support of the government. Since 2010, the state and the government have promulgated many policies to promote the development of the smart home industry. In 2010, the State Council passed the "Decision on Accelerating the Fostering and Development of Strategic Emerging Industries" issued by the State Council in 2011, and the Ministry of Industry and Information Technology promulgated the "Twelfth Five-Year Plan for Development of the Internet of Things" in 2011, explicitly including smart home in the demonstration projects of 9 major key areas. It is precisely because of state support that smart home as a new industry can be developed so rapidly, largely to help improve the technology and increase the popularity of smart home.

3.2.3. Business competition. In other countries, Apple, Google and other leading enterprises have long been put into the smart home industry. In China, Alibaba, 360, Xiaomi and Jingdong started to think of smart home as an important area for future development. The emergence of many competitors in some ways to promote the development of the industry has a positive lead role.

4. The overall design of the system

It takes Andriod system as a platform, the system includes CPU, WiFi module, a smart phone as a carrier control terminal and intelligent control system. Intelligent control system includes home lighting intelligent control system, home appliances intelligent control system, audio-visual equipment intelligent control system and video real-time monitoring system, and finally by Andriod system smart phones through WiFi network to achieve the purpose of controlling home appliances.

The function of this system is to send the control command to the CPU by the smart phone through the WiFi network. The processor receives the command and parses it to remotely set the door, air conditioner, monitoring probe and light. Users can also use the smart phone to gather real-time data sent by the central processing unit.

Hardware platform is an important part of determining system performance and stability. This article chooses STM32 core controller as the central processor, QUALCOMM's QCA9531 chip as the WiFi module, the smart phone based on Andriod system as the control terminal.

The control terminal transmits the information to the WiFi module connected to the same network via WiFi. The WiFi module transmits the received signal to the CPU through the serial port, and the CPU realizes control over the home appliance according to the received instructions. The overall structure of the system shown in Figure 1.

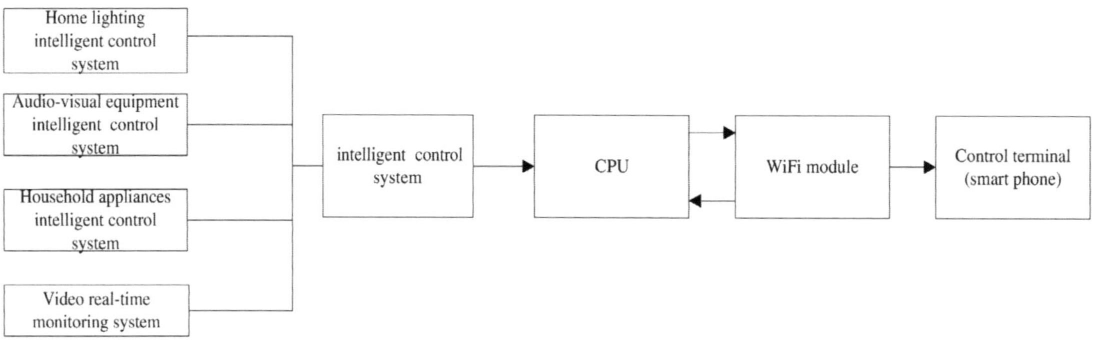

Figure 1. Overall structure diagram of the system

5. The hardware design of System

STM32F107 processor module's main role is to communicate with the host computer to receive real-time operation from the mobile phone side instructions, and then analyze the operation of instructions and perform the appropriate work; The OpenWrt firmware is embedded in the QCA9531WiFi wireless communication module, and the data forwarding and transmission of the QCA9531WiFi wireless communication module is realized by configuring the network serial port function and the push function of the OpenWrt system. The STM32F107 processor mainly realizes the data communication with the WiFi wireless communication module through the serial port and provides 5V power for the WiFi wireless module.

5.1. STM32 processor module

The STM32F107 microcontroller uses the Cortex-M3 core with CPU speeds up to 72 MHz. They target applications requiring connectivity and real-time performance with 64-256KB of on-chip Flash memory, 64KB of SRAM and 14 communications interfaces. The STM32F107 Interconnect has a built-in USB OTG full-speed (12Mb / s) device, host and OTG mode control module. The biggest highlight of the STM32F107VCT chip is the integration of most of the mainstream peripheral networks, CAN, USB, UART, motor control and so on.

Compared to using more STM32F103VB series, the STM32F107VCT series has an interconnection interface and more internal resources, the F107 increases the number of the IEEE Ethernet interface, it has two IIS audio interface, and all 64kb SRAM cache.

5.2. WiFi wireless communication module

The WiFi wireless communication module in this system adopts Qualcomm QCA9531 chip. QCA9531 frequency will be higher, up to 650MHZ . QCA9531 supports MINIPCIE interface and can be extended to 5G dual-band and LTE data module applications, QCA9531 increases the number of external interfaces and increases memory that is up to 128MB, QCA9531 improves performance and expand memory, and optimizes for radio frequency. Compared with the AR9341, there will be no restricted situation, and it has more associated users, supporting the concurrent number of up to 50 users, with a measurement of 150 people. QCA9531's block diagram shown in Figure 2.

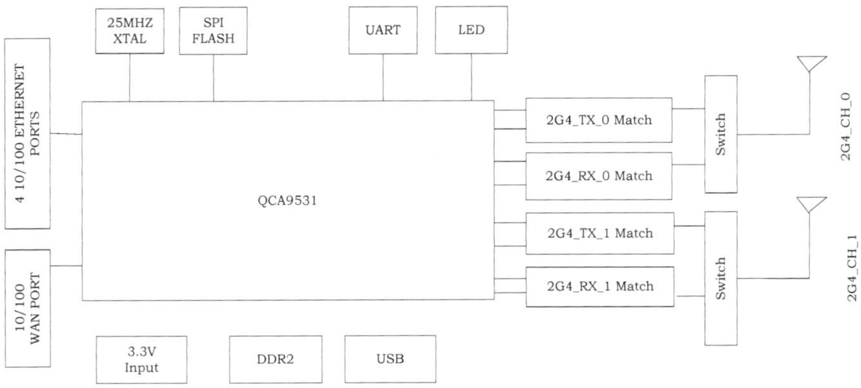

Figure 2. QCA9531 module diagram

6. The software design of System

The software design of the system is based on the Andriod system above the Linux system. Andriod operating system with high degree of freedom, openness, wide range of applications, and has great potential for development, it is also very competitive in the future. This article chooses Andriod operating system as a platform to develop the intelligent home control system. By installing the APP interface on the smart phone, the user can realize the switch control of the home appliance, the centralized control of the home lighting, and real-time monitoring of the home situation, it not only can control the electrical equipment in the house centrally, but also can realize the remote management, reduce the potential safety hazard and apply the modern advanced technology to people's life truly.

APP software interface is mainly consisted by welcome interface, login interface, the main interface. User interface design is shown in Figure 3. Firstly, the user needs to register an account in the APP, set a password, and log in to the main interface of the management through the registered account to ensure the privacy of the user. The main interface contains settings interface, you can set the switch tone, the interface style and so on. It can increase the users' experience and personalization. Home control interface has four sub-modules: smart home appliances control interface, smart home lighting control interface, audio and video equipment control interface and video surveillance interface.

IOP Conf. Series: Journal of Physics: Conf. Series **1061** (2018) 012010 doi:10.1088/1742-6596/1061/1/012010

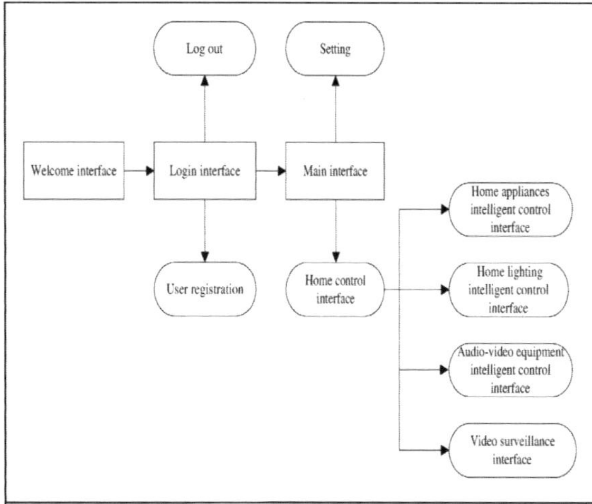

Figure 3. Design of user interface operation

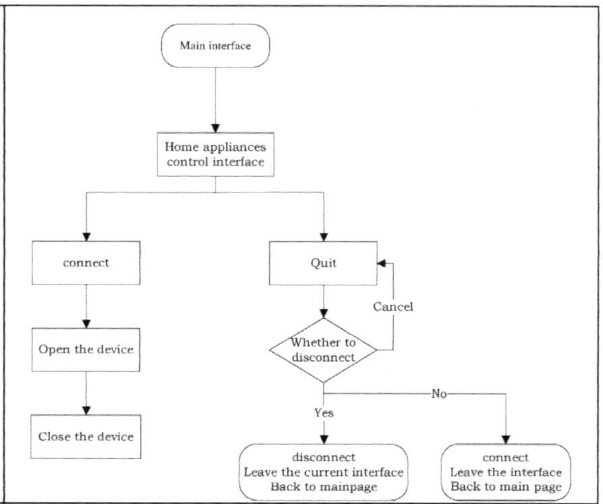

Figure 4. Subsystem workflow

This article uses MATLAB GUI user interface to simulate the design of the interface. For example, take home appliances control interface to introduce its subroutine flow chart shown in Figure 4. There are two "connect" and "exit connection" control buttons respectively, click the connect button to connect the device, after the success of the connection ,it can choose to open and close the home appliances; click the exit button can choose to exit whether to disconnect, if so, then Disconnect the device and exit the current interface and return to the main interface; If not, do not disconnect the device and return to the main interface to control the other sub-modules; If canceled, it stay in the current interface. The main interface is shown in Figure 5, the subsystem interface is shown in Figure 6.

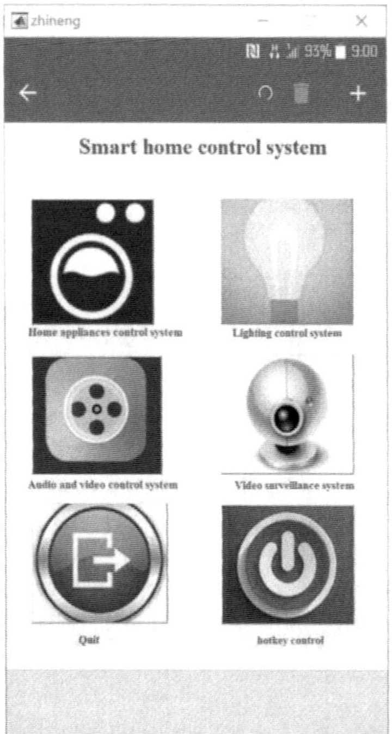

Figure 5. The main interface of Smart home control system

Figure 6. Home appliances control interface

Conclusion

This article firstly introduces the WiFi technology principle and the development overview of the smart home, Secondly, the design of hardware and software of this system is described. The STM32F107VCT chips and QCA9531 is used to control the home system through Andriod platform. Finally, the APP of Matlab software is designed based on GUI of Matlab.

The system has the following advantages:

- The system structure is simple, the user interface is concise, cost-effective and easy to operate.
- Choose STM32 processor to be the central processor, the selected STM32F107VCT series is an Internet-connected products for Internet of things, it is cost-effective. Choose QCA9531 as WiFi wireless communication module, QCA9531 not only increases the number of interfaces and has more memory, but also increases the number of concurrent users, associates more users, optimizes the overall performance.
- The system uses Matlab GUI interface for app design, it can realize the appropriate functions, users can easily control smart home appliances to realize the function of equipment.

References

[1] F. Liu and H. Zhao, "The Design of WIFI-Based Smart Home Communication Hardware Adapter," *2015 Fifth International Conference on Instrumentation and Measurement, Computer, Communication and Control (IMCCC)*, Qinhuangdao, 2015, pp. 1193-1197.

[2] A Review of Smart Homes-Past, Present, and Future. Muhammad Raisul Alam,Mamun Bin Ibne Reaz,Mohd Alauddin Mohd Ali. IEEE TRANSACTIONS ON SYSTEMS MAN AND CYBERNETICS PART C-APPLICATIONS AND REVIEWS . 2012

[3] Y. Wenbo, W. Quanyu and G. Zhenwei, "Smart home implementation based on Internet and WiFi technology," *2015 34th Chinese Control Conference (CCC)*, Hangzhou, 2015, pp. 9072-9077.

[4] A. Nacer, B. Marhic and L. Delahoche, "Smart Home, Smart HEMS, Smart heating: An overview of the latest products and trends," *2017 6th International Conference on Systems and Control (ICSC)*, Batna, 2017, pp. 90-95.

[5] H. Jiang, B. Liu and C. W. Chen, "Performance analysis for ZigBee under WiFi interference in smart home,"*2017 IEEE International Conference on Communications (ICC)*, Paris, 2017, pp. 1-6.

IOP Conf. Series: Journal of Physics: Conf. Series **1061** (2018) 012011 doi:10.1088/1742-6596/1061/1/012011

Computer Modeling of the Nonstationary State of Saturation of A Membrane in the Form of Taking Into Account Relaxation Times

Zhanat Umarova[1], Saule Botayeva[1], Zhalgasbek Iztayev[1], Aigul Imanbayeva[1] and Asylzat Kopzhasarova[1]

[1]Department of Information systems and modeling, South Kazakhstan State University, Information technology and energetic scientific school, Tauke khan avenue, 5, Shymkent, Kazakhstan

*Email: Zhanat-u@mail.ru

Abstract. The paper reviews a model is developed for the nonstationary state of saturation of a membrane. In the main part of the paper, transfer equations with cross effects such as the Soret effect and thermal diffusion are derived. On the basis of these equations, cross effects, heat transfer and mass transfer equations in the conjugate form are obtained. As a result, equations are obtained for component fluxes in the transition period, taking into account the relaxation time, where the aspect of membrane nonstationarity is associated with transient processes in establishing the state of stationary saturation of the membrane. Moreover, it was shown that the concentration of the dissolved substance near the membrane does not exceed the equilibrium concentration, which avoids clogging of pores with polymeric molecules. However, the apparent separate porosities for the dissolved component and the solvent in the "slip" porosity zone depend on the saturation of the membrane pores by the filtering components. All this is described in the discussions of the paper. In the results, the computer model of non-stationary saturation of the membrane with the application of the Chemkad application program allowed to obtain graphs of the dependence of the non-stationary saturation state of the membrane on various gases in the porosity zone, in particular, gases such as ethanol and methane are considered.

1. Introduction

Numerous phenomena of diffusion, mass transfer and migration vary significantly from a physical point of view, but can be described by a single mathematical apparatus. This apparatus has a complex nature, including such directions as the statistics of forced or random walks, the phenomenological description of diffusion by partial differential equations and elements of nonequilibrium thermodynamics.

The mathematical apparatus of diffusion is quite complex, it is not inferior in complexity to the apparatus of quantum mechanics, and in some cases exceeds it. Diffusion uses statistical distributions that do not have finite variance and higher moments. Some equations have no solution at all.

In practice, the analogy between diffusion and thermal conductivity is often used. Mathematically, Fick's laws are analogous to the equations of Fourier thermal conductivity. This analogy is based on the general patterns of irreversible processes of redistribution of the intensive parameters of the state (concentration, temperature, pressure, etc.) between different parts of a system as it tends to thermodynamic equilibrium. [1]

In traditional membrane technology, a mixture of gases is separated only from stationary conditions. The partition factor is defined as αss = PA / PB. Under stationary conditions, by definition, a mixture of gases A and B for which PA = PB cannot be separated.

However, such a mixture, can be separated if the separation is conducted in the non-stationary permeability mode. In this case, it is necessary to take into account both the relaxation times of the thermal and diffusion phenomena.

Therefore, it should be written transfer equations with cross effects such as the Soret effect and thermal diffusion.[2]

2. The Main Part

In order to take into account to cross effects, the heat transfer and mass transfer equations must be written in conjugated form. It can be done as follows:

$$\frac{\partial N_m}{\partial t} = -N_m \tau_m^{-1} + N_h \tau_x^{-1} \tag{1}$$

$$\frac{\partial N_h}{\partial t} = -N_m \tau_x^{-1} + N_h \tau_h^{-1} \tag{2}$$

Herewith, τ_x can be determined as the perturbation time by analogy with the relaxation time. The indexes m and h mean mass and heat, respectively.[3] And it is important that in the quasiequilibrium approximation of the system the condition for the damping of perturbations be fulfilled [4]:

$$\tau_x > \sqrt{\tau_m \tau_h} \tag{3}$$

The obtained solutions show that the first terms in the equations are related to the direct flows, and the second ones are the flows associated with the cross flows. [5] Then the expressions for the corresponding relaxation kernels of transfer are obtained in the forms [6]:

$$N_{mm} = \frac{\eta_m}{\lambda_2 - \lambda_1}\left[\left(\lambda_2 + \frac{1}{\tau_m}\right)\exp(\lambda_1 s) - \left(\lambda_1 + \frac{1}{\tau_m}\right)\exp(\lambda_2 s)\right] \tag{4}$$

$$N_{hh} = \frac{\eta_h}{\lambda_2 - \lambda_1}\left[\left(\lambda_2 + \frac{1}{\tau_h}\right)\exp(\lambda_1 s) - \left(\lambda_1 + \frac{1}{\tau_h}\right)\exp(\lambda_2 s)\right] \tag{5}$$

$$N_{mh} = N_{hm} = \frac{\eta_x}{\tau_x(\lambda_2 - \lambda_1)}\left[\exp(\lambda_1 s) - \exp(\lambda_2 s)\right] \tag{6}$$

Since the fluxes of both components in a two-component system are linearly interrelated and, taking into account the low concentration of the diffusing impurity,[7] we can adopt the following expressions of the fluxes:

$$J_m = \int_0^t N_{mm} \beta \nabla \nu(R,t_1) dt_1 + \int_0^t N_{mh} \nabla \beta(R,t_1) dt_1 \quad (7)$$

$$J_h = \int_0^t N_{hm} \beta \nabla \nu(R,t_1) dt_1 + \int_0^t N_{hh} \nabla \beta(R,t_1) dt_1 \quad (8)$$

Hence the equations for the streams take the form:

$$\frac{\partial^2 J_m}{\partial t^2} = (D_{1-2}) \frac{\partial J_m}{\partial t} - D_{12} J_m - \eta_m \beta \nabla \nu \left(C + \frac{1}{\tau_m} \right) + \frac{\eta_x}{\tau_x} \nabla \beta + \eta_m \frac{\partial}{\partial t} (\beta \nabla \nu) \quad (9)$$

$$\frac{\partial^2 J_h}{\partial t^2} = D_{1-2} \frac{\partial J_h}{\partial t} - D_{12} J_h - \eta_h \nabla \beta \left(C + \frac{1}{\tau_h} \right) + \frac{\eta_x}{\tau_x} \beta \nabla \nu + \eta_h \frac{\partial}{\partial t} (\nabla \beta) \quad (10)$$

The laws [8] of conservation of mass and energy can be written as:

$$\frac{\partial c}{\partial t} + \nabla J_m = 0 \quad (11)$$

$$\frac{\partial E}{\partial t} + \nabla J_h = 0 \quad (12)$$

As a result, the transfer equations become:

$$\frac{\partial^3 c}{\partial t^3} - \left(\frac{1}{\tau_m} + \frac{1}{\tau_h} \right) \frac{\partial^2 c}{\partial t^2} + \left(\frac{1}{\tau_m \tau_h} - \frac{1}{\tau_x^2} \right) \frac{\partial c}{\partial t} + \frac{\eta_m}{\tau_h} \nabla(\beta \nabla \nu) + \frac{\eta_x}{\tau_x} \Delta \beta + \eta_m \frac{\partial}{\partial t} \nabla(\beta \nabla \nu) = 0 \quad (13)$$

$$\frac{\partial^3 E}{\partial t^3} - \left(\frac{1}{\tau_m} + \frac{1}{\tau_h} \right) \frac{\partial^2 E}{\partial t^2} + \left(\frac{1}{\tau_m \tau_h} - \frac{1}{\tau_x^2} \right) \frac{\partial E}{\partial t} + \frac{\eta_x}{\tau_x} \nabla(\beta \nabla \nu) + \frac{\eta_h}{\tau_m} \Delta \beta + \eta_h \frac{\partial}{\partial t} (\Delta \beta) = 0 \quad (14)$$

3. Discussion

Consequently, the aspect of membrane nonstationarity is associated with transient processes of establishing the state of stationary saturation of the membrane. [9]

The apparent separate porosities for the dissolved component and the solvent in the "slip" porosity zones to depend on the saturation of the membrane pores by the filtering components and, accordingly,

$$\varepsilon_1 = \varepsilon S_1, \quad \varepsilon_2 = \varepsilon S_2, \quad (15)$$

Where, as a hypothesis, it can be assumed that for an unsteady process at any time, an equality similar to that for a stationary process is observed:

$$S_1 + S_2 = \frac{\varepsilon_P}{\varepsilon} \quad (16)$$

With steady-state ultrafiltration in the flow regime, no dissolved component is accumulated in the membrane in the "slip" stream filtration zone. [10] This condition is written as:

$$C_P = \frac{V_2}{V_1 + V_2} = f(z) \tag{17}$$

$$\frac{\partial C_P}{\partial y} = 0 \tag{18}$$

In the transition period, we can approximate the function $C_p(t)$ as:

$$\frac{\partial C_P}{\partial t} = \frac{\partial}{\partial t}\left(\frac{V_2}{V_1 + V_2}\right) = f(z,t) \tag{19}$$

Furthermore

$$J_P = \frac{\overline{C_P V_1}}{1 - \overline{\overline{C_P}}} \tag{20}$$

$$\varepsilon_S = 1 - \varepsilon_P \tag{21}$$

4. Results

Thus, we have obtained a nonlocal description of the separation process, taking into account the long-term effects of memorization in the system of globules and the non-commutability of the system.[11] The drift intensity at times steps depend on the history of the process and is determined by the kinetic characteristics at all preceding moments $\tau_1 \le \tau$. [12]

Figure 1 shows a characteristic form of the dynamics of the cross component of the relaxation transfer nucleus.

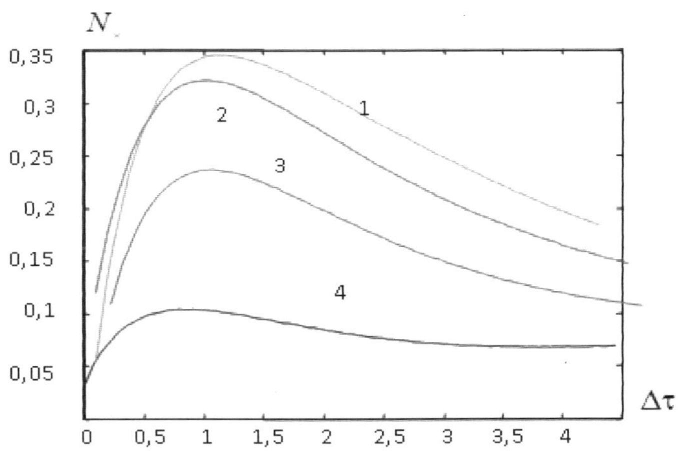

Figure 1. The dynamic of the cross relaxation perturbation

Moreover, the effective diffusion coefficient changes with time when the structure of the kinetic zones changes during the flow of the mixture through the membrane. [13]

The model predicts an increase in the effective diffusion coefficient in time and the displacement of the fractal zone in the membrane.

Applying a package of Chemcad application programs, we get the following dependency curves for ethanol and methane at the porosity zone (Figure 2, Figure 3).

Library Ideal Gas Heat Capacity. Ethanol

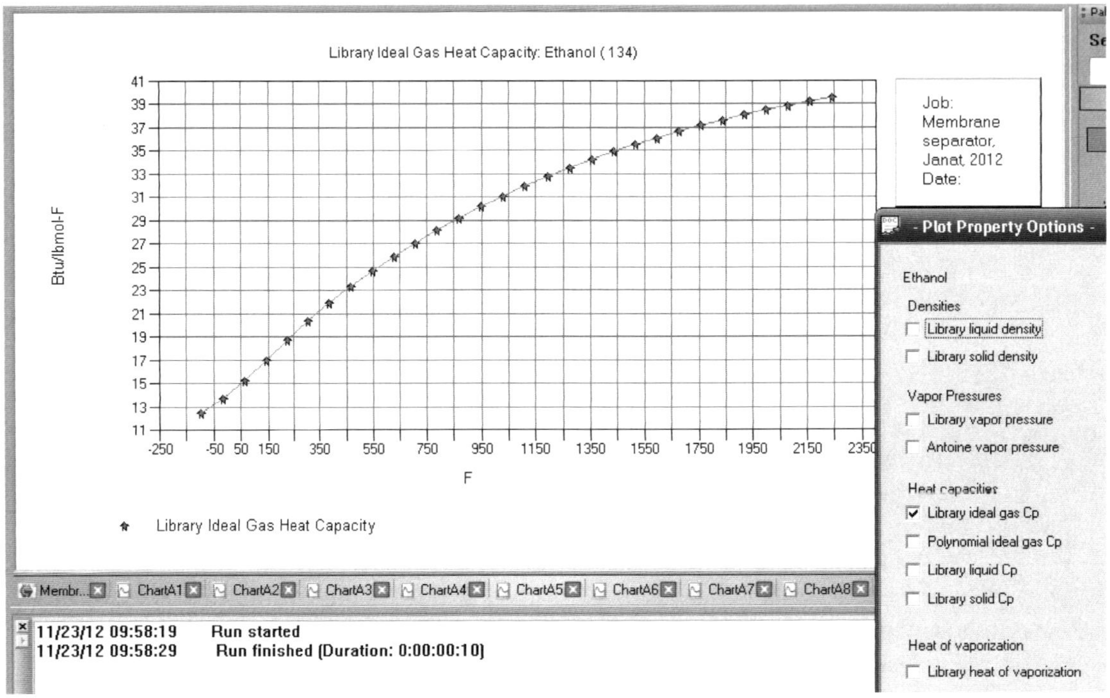

Figure 2. Diagram for the Ethanol component

Library Ideal Gas Heat Capacity. Methane

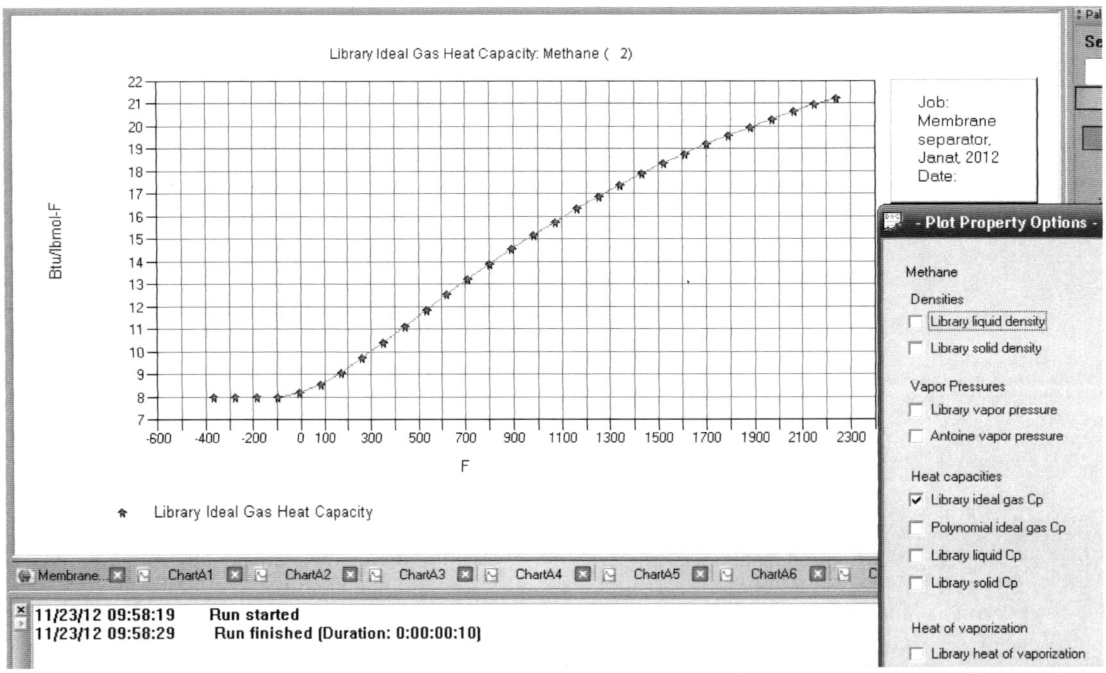

Figure 3. Diagram for the Methane component

IOP Conf. Series: Journal of Physics: Conf. Series **1061** (2018) 012011 doi:10.1088/1742-6596/1061/1/012011

5. Conclusion

Only the solvent moves through this channel system. Moreover, supposing that the concentration of the dissolved substance near the membrane does not exceed the equilibrium concentration, clogging of the pores with polymer molecules does not occur.
Based on this model, we obtain equations for the flow of components in the transition period.

References

[1] Bekman I 2016 Matematika diffuzii Study guideline Moscow
[2] Timashev S 1988 Fiziko-khimiya membrannykh protsessov. Moscow
[3] Ovchinnikov A, Timashev S, Belyj A 1986 Kinetika diffuzionno-kontroliruemykh khimicheskikh protsessov Moscow
[4] Yasminov A 1978 Obrabotka vody obratnym osmosom i ultrafiltratsiej Moscow
[5] Nakagaki M 1991 Fizicheskaya khimiya membrane Moscow
[6] Bekman I 2015 Neobychnye membrannye protsessy: nestatsionarnye rezhimy, negomogennye i dvizhushhiesya membrany
[7] Snyder M, Vlachos D, Katsoulakis M 2003 Mesoscopic modeling of transport and reaction in microporous crystalline membranes// Chem. Eng. Science, V. 58
[8] Horntrop D, Katsoulakis M, Vlachos D 2001 Spectral methods for mesoscopic models of pattern formation// Jour. of Comp. Physics p 173
[9] Vlachos D, Katsoulakis M 2000 Derivation and validation of mesoscopic theories for diffusion of interacting molecules// Phys. Review Letters, V.85(18)
[10] Brener A, Muratov A, Tashimov L 2004 Non-linear model of time-dependent relaxation cores for the systems with cross transfer effects// Adv. Comp. Methods in Heat Trans., VIII,WIT Press
[11] Reese J, Raimondeau S, Vlachos D 2001 Monte Carlo algorithms for complex surface reaction mechanisms: Efficiency and accuracy// Jour. of Comp. Physics 173(10)
[12] S.Pereira Nunes and K-V Peinemann 2006 Membrane Technology in the Chemical Industry Wiley-VCH
[13] Richard W 2004 Baker,Membrane Technology and Applications Wiley

IOP Conf. Series: Journal of Physics: Conf. Series **1061** (2018) 012012 doi:10.1088/1742-6596/1061/1/012012

Time Series Prediction and Anomaly Detection of Light Curve Using LSTM Neural Network

Runtian Zhang[1, a] **and Qian Zou**[1,b]

[1]Department of Information, Beijing University of Technology, Beijing, China

*[a]rtzhang9317@126.com; [b]zouqian@emails.bjut.edu.cn

Abstract. Ground-based Wide-angle Camera array (GWAC) is a short-time survey telescope which can produce images every 15 seconds for more than 30,000 stars. Light curve is generated from star image with a series of processing. Research on light curve is a new task in time domain astronomy which can detect anomaly astronomical events. We explore a series prediction model of LSTM neural network for light curve prediction. Through model training and validation we obtain the optimal structure. Then we predict one time-step ahead light luminance for test star. We evaluate the performance of model by calculating prediction error. Anomaly detection mechanism is based on prediction error. Experimental results based on real light curve data demonstrates that our model is promising in light curve prediction and anomaly detection.

1. Introduction

With the rapid development of astronomical observed capability and data processing technology, an enormous amount of data is increasing exponentially. Monitoring data plays a critical role of astronomical studies [1], such as discovery of early supernova eruption, fast response observation of gamma-ray burst and light curve observation [2]. Light curve research is a new task in time domain astronomy [3]. Large datasets are generated by sky survey and transformed to time-series data. GWAC (Ground-based Wide-angle Camera array) is the important equipment in Space Multi-band Variable Objects Monitor mission [4]-[6], which has the capability of providing astronomical data in short time-scale. SVOM aims to study gamma-ray burst and this phenomenon leads to a sudden abnormal of light luminance. Research on anomaly detection of real-time light luminance (light curve) is significant which is the purpose of this paper.

Traditional time series anomaly detection is based on statistical approach. For instance, an Auto Regressive Moving Average (ARMA) based model [7] simulates stationary time sequences over a time window to detect anomaly values. The length of window is pre-determined and results are largely depends on it. The Kalman filter [8]-[9] is an algorithm for the optimal estimation of the system states It is considered as a filtering process due to the observation data includes the influence of noise and interference in the system. It is obvious that anomaly detection is generally based on comparing observation and its predicted value at the same time step. RNN(Recurrent Neural Networks)is a neural network that captures the dynamic information in serialized data by periodically connecting hidden layers. LSTM (Long Short-Term Memory neural network) is an improved recurrent neural network. It overcomes gradient vanished problems during long time data training process by using memory cell. [9] This characteristic makes LSTM networks applicable to time series problem.

It's a worthwhile study that combining this advantage of predicting time series with anomaly detection based on time prediction. As in [10], it implements anomaly captured by training normal sequence and anomaly sequence. However, the difficulty of our purpose is that light curve is streaming data based on star luminance which has no periodicity and reference anomaly sequence. And the observed data is obtained every 15 seconds, which means anomalies need to be captured within 15 seconds for each spot. To the best of our knowledge, it is the first time that the LSTM neural network is used on astronomical time problem.

The rest of the paper is organized as follows. Section 2 describes the detail of light curve dataset and real-time series acquisition mechanism. Section 3 introduces the LSTM neural network model for prediction and data pre-processing method. Section 4 presents our experiment based on the LSTM neural network using light curve dataset, introduces anomaly detection mechanism, analyses the performance of our model using results. Section 5 concludes this paper.

2. Light curve dataset

Light curve is a series of light luminance data over time. The dataset is generated from GWAC which called mini-GWAC. It is obtained from the National Astronomical Observatories of China. The detail of our dataset is as following:

- GWAC can produce images every 15 seconds for more than 30,000 stars. After a series of processing, such as quality control, source extraction, flux calibration, source association, each time-step light luminance value generated.
- The amount of each star-captured data is almost 900 per day. GWAC is in trial operation and our dataset contains three days of observations.
- We verify the result by using the data of object with catalog ID of 3348.

In present, we only have three days data for evaluating our approach. But after the formal operation of GWAC, a huge amount of light luminance data generated every day. With those history data, our experiment will be more effective in the future.

3. Methodology

3.1 The structure of LSTM

LSTM neural network has a great contribution to solving real-world problems in many fileds. Comparing with RNN, the most significant improvement of LSTM is having a complex structure called memory cell in hidden layer. As is shown in Figure 1, x_t is the input layer sequence at time t, A is a hidden layer with memory cell, h_t is the hidden layer output sequence at time t. In the process of data transfer in the hidden layer, C_t is a special information which contains the state of cell. The main purpose of the transition is to update the cell state and output h_t at time t. The specific transition in LSTM cell is shown in Figure 2. Each of the yellow block is neural network layer which reads data from

h_{t-1} and x_t. Blue blocks are calculative units which are designed for vector operation.

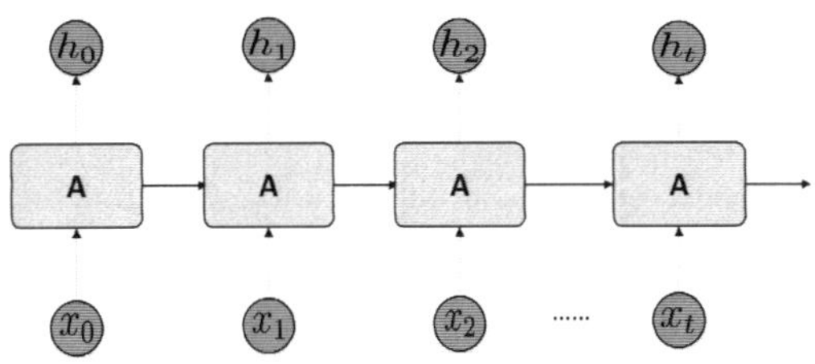

Figure 1. The structure of LSTM with series prediction

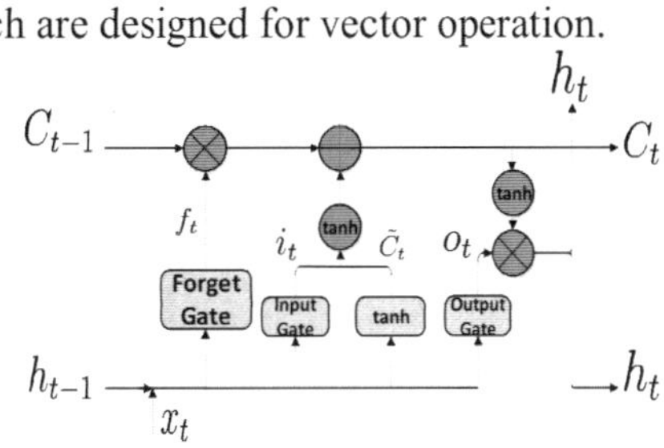

Figure 2. The structure of LSTM cell

We firstly decide what information need to be discarded which is generated by forget gate. It outputs results from 0 to 1 for each number in cell state C_{t-1}. "0" is represented "completely discarded", analogously, "1" is represented "completely reservation". Next step is the decision of choosing new information of cell state. The first part is input gate which outputs i_t that need to be updated. "tanh" layer creates a new vector $\tilde{C_t}$. Finally, the output gate's state is o_t. Both C_{t-1} and h_{t-1} are well transmitted under the structure of LSTM cell. Equations of the transmission process is shown below:

- Forget gate's state:

$$f_t = \sigma(W_f \cdot [h_{t-1}, x_t] + b_f) \quad (1)$$

- Input gate's state:

$$i_t = \sigma(W_i \cdot [h_{t-1}, x_t] + b_i) \quad (2)$$

$$\tilde{C_t} = tanh(W_C \cdot [h_{t-1}, x_t] + b_C) \quad \text{- (3)}$$

- Updated C_t:

$$C_t = f_t * C_{t-1} + i_t * \tilde{C_t} \quad (4)$$

- Output gate's state:

$$o_t - \sigma(W_o[h_{t-1}, x_t] + b_o) \quad (5)$$

- Output h_t:

$$h_t = o_t * tanh(C_t) \quad (6)$$

For parameters in equations (1)-(6), σ represents the sigmoid activation function $(1 + exp^{-x})^{-1}$, tanh means the hyperbolic tangent activation function $(exp^x - exp^{-x}) * (exp^x + exp^{-x})^{-1}$,$W_f$, W_i, W_C, W_o are the weight matrices for both x_t and h_{t-1} in each gate, b_f, b_i, b_C, b_o are bias of each gate.

3.2 Stationary

The result of LSTM model will be more flexible if time series is stationary. There is a huge fluctuation of light luminance if gamma-ray burst is happening. Besides, the history data is merely 3 days. It's uncertain if history data has a long-term trend. Considering all of these factors, it's essential to make sequence stationary. A common method for stationary is "differencing". It's the observation that previous time step (t-1) is subtracted from current time step t. The original history data $X = \{X_1, X_2, \cdots, X_t\}$ is changed. $\tilde{X} = \{X_1 - X_0, X_2 - X_1, \cdots, X_t - X_{t-1}\}$ is the new format of historical data. (We assume that X_0 equals to X_1).

4. Experiments

We construct our LSTM model for real-time light curve series from mini-GWAC dataset. The amount of history data for ID 3348 is 2777. We divide it into three parts: training set (75%) as historical data, validation set (15%) for evaluating predicted error and test set (15%) for real-time prediction. In order to simulate capturing anomalies from real-time generated data, test set is considered as the real-time data which is produced by mini-GWAC every 15 seconds one by one. It indicates that our model focus on predicting one time period value.

4.1. Model training

We employ differencing method for data pre-processing. Processed datasets are shown below:

$$X_{train} = \{X_{h1} - X_{h0}, X_{h2} - X_{h1}, \cdots, X_{lenH} - X_{lenH-1}\}, \quad (7)$$

$$X_{val} = \{X_{v1} - X_{v0}, X_{v2} - X_{v1}, \cdots, X_{lenV} - X_{lenV-1}\}, \quad (8)$$

$$X_{test} = \{X_{t1} - X_{t0}, X_{t2} - X_{t1}, \cdots, X_{lenT} - X_{lenT-1}\}, \quad (9)$$

$$X_{origin} = \{X_{h1}, X_{h2}, \cdots, X_{lenH}, X_{v1}, X_{v2}, \cdots, X_{lenV}, X_{t1}, X_{t2}, \cdots, X_{lenT}, \} \quad (10)$$

WhereX_{origin} is original dataset and all of the divided parts are assigned in order of the original timestamp. For instance, $X_{v0} = X_{lenH}$ and $X_{t0} = X_{lenV}$.

The dimension of input layer and output layer is 1 due to the number of light luminance at each time. In series prediction LSTM neural network, there are many factors affecting the performance of model. We evaluate performance by comparing results of different number of hidden units and epoch time. The number of hidden units is ranging from 1 to 5 and epoch time is ranging from 1 to 100. Loss function of the model is "Mean_Squared_Error" which closely matches MSE that is used for calculating predicted error.

As is shown in Figure 3, blue lines are training error and red lines are validation error. Training set error tends to be stable through less than 10 epochs and it is much better than validation error. The number of cells has little effect on the performance. We construct the model with 5 hidden layer units and 100 epoch times of training process for stable result.

4.2. Model Evaluation

After compiling our model, test set predicts the value of next time period by calculating parameters of the model. Then prediction results are converted from differential form to origin form. We measure the prediction error of both train set and test set by mean square error (MSE),

$$MSE = \frac{1}{N}\sum_{i=1}^{N}(X_t - \tilde{X}_t)^2 \tag{11}$$

Where N is the number of sequence, X_t and $\tilde{X}_t$ are observed value and predicted value at time step t. We repeat 10 times of the whole prediction process. Statistical result is shown in Table 1. The result of Table 1. Indicates that prediction of train set is much better than test set. It is reasonable that light luminance of test set has a huge fluctuation comparing with train set. The real light curve of ID "3348" is shown in Figure 4.

4.3. Anomaly detection mechanism

Anomaly detection mechanism is based on MSE of both train set and test set. It is used for instant warning if the prediction of current observation Considering the prediction of train set is much better than test set, an experimental ratio is provided to scale error between train set and test set. We measure the anomalies slot by following rule, Ano_{slot} is determined at time t if

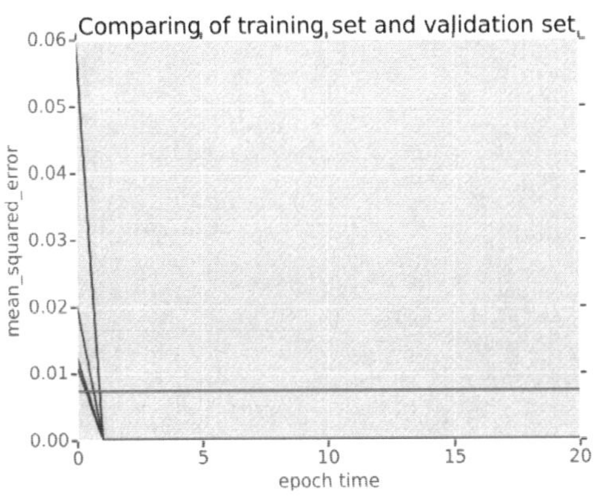

Figure 3. "mean_squared_error" of train set and validation set

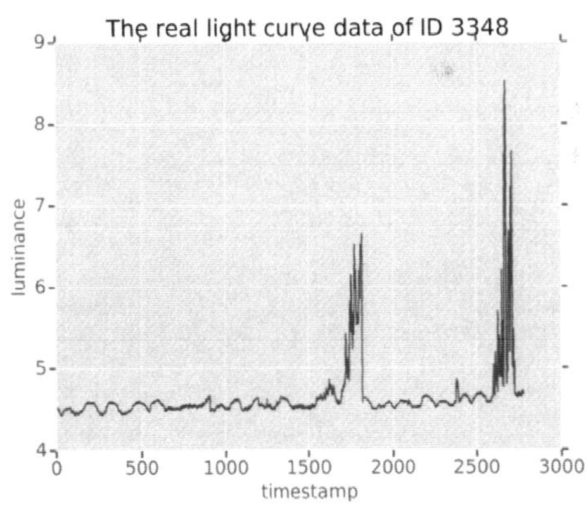

Figure 4. The real light curve data of ID 3348

Table 1. Statistical result of MSE of train set and test set

	Max	min	mean	variance
Train set	0.013515	0.010378	0.011908	1.107e-6
Test set	0.572562	0.199559	0.323205	0.000647

$$(Predicted_t - Observed_t)^2 * R > MSE_{maxtrain} \quad (12)$$

Where $MSE_{maxtrain}$ is the max MSE of train set, R is the ratio that is set as 0.3. The mechanism is tested based on both real data and manually added anomalies data. It is significant that testing our approach under varies situation. We manually add random data ranged from 2 to 8, similarly, a sin wave is added on another situation. Figure 5. shows the result with added anomalies data. Blue line is represented both training set and validation set, black dots are test data, red dots are anomalies slots. Figure 6. shows the result with real test set. Our approach finds anomalies value under sudden fluctuation. This is sufficient for the purpose of instant warning. It also generates some false results and miss some anomalies slot such as waved situation. We strongly believe our method can be improved if we have more historical data and some known astronomical principle.

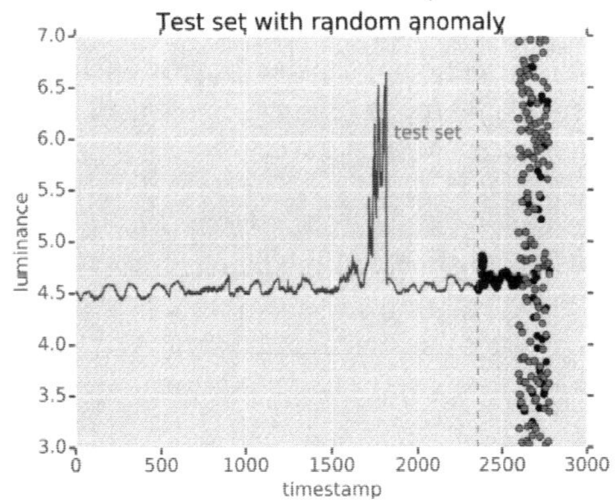

(a)

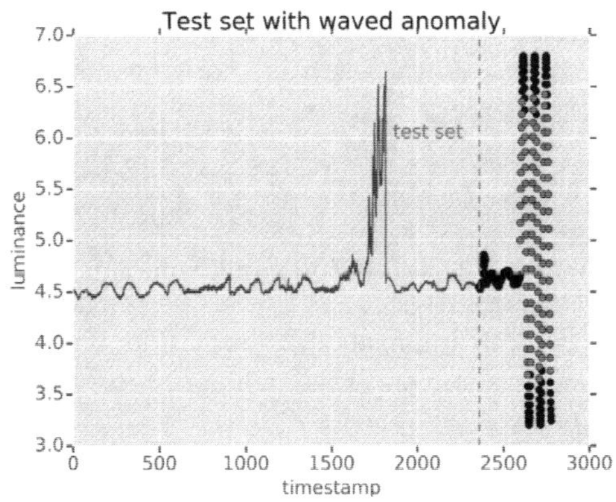

(b)

Figure 5. Test set with manually added anomalies data. (a) is represented results with random anomalies. (b) is represented results with waved anomalies.

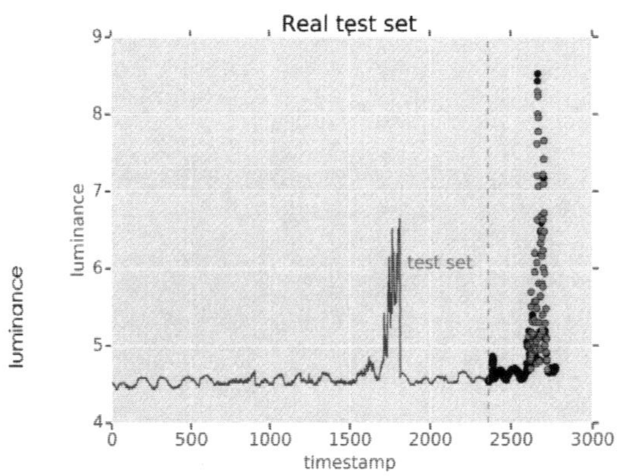

Figure 6. Real test set with anomaly

5. Conclusion

We explorer a LSTM neural network model for light curve time series predictions. Employing light curve data provided by mini-GWAC dataset. Then we propose the anomaly detection mechanism based on prediction error. This mechanism is sufficient for real test set and test set of manually added anomalies slot. In the future , we plan to evaluate our model on GWAC system if is ready to operate on line.

IOP Conf. Series: Journal of Physics: Conf. Series **1061** (2018) 012012 doi:10.1088/1742-6596/1061/1/012012

Reference

[1] S. Vaughan, "Random time series in astronomy," Philosophical Transactions of the Royal Society of London, vol. 371, no.1984, 2012.

[2] Ann M, "The Swift gamma-ray burst explorer Burst Alert Telescope," Nuclear Science Symposium and Medical Imaging Conference, vol. 4, pp 2382-2386, 2002.

[3] Zhang Li, Bian Xia, Mei Dong-Cheng; "Gamma-ray light curve and phase-resolved spectra from Geminga pulsar," vol. 10, no. 7, pp 662-665, 2001.

[4] channe S, Paul J, Wei J, "The future gamma-ray burst mission SVOM," Proceedings of Science, E-ISSN: 18248039, 2009.

[5] Wan Meng, Wu Chao, Ying Zhang, "A Pre-research on GWAC Massive Catalog Data Storage and Processing System," Astronomical Research and Technology, no. 3, pp. 373-381, 2016.

[6] Wan Meng, "Column store for gwac: a High-cadence, High-density, Large-scale astronomical light curve pipeline and distributed Shared-nothing database," Publications of the Astronomical Society of the Pacific, no. 969, 2016.

[7] Aliat Billel, Hamdi Faycal, "On Markov-switching periodic ARMA models," Communication in Statistics, vol. 47, no. 2, pp 344-364, 2017.

[8] Janjic Tijana, Nerger Lars, "On domain localization in ensemble-based Kalman filter algoithms," Monthly Weather Review, American Meteorological Society,vol. 139,no. 7, pp 2046-2060,2011.

[9] Han Fei, Dong Hongli, Wang Zidong, "Improved Tobit Kalman filtering for systems with random parameters via conditional expectation," Signal Processing, vol. 147, pp 35-45, 2018.

[10] Malhotra Pankaj, Vig lovekesh, Shroff Gautam, "Long Short Term Memory networks for anomaly detection in time series," 23rd European Symposium on Artificial Neural Networks, pp 89-94, 2015.

IOP Conf. Series: Journal of Physics: Conf. Series **1061** (2018) 012013 doi:10.1088/1742-6596/1061/1/012013

Observer Design for Switched Nonlinear System with Both Unknown Inputs and Measurement Noise

Fanglai Zhu[1,*], Jiancheng Zhang[1,2] and Fengning Wang[1]

[1]College of Electronics and Information Engineering, Tongji University, Shanghai, 201804, China
[2]School of Science, Jiangnan University, Wuxi, 214122, China

*zhufanglai@tongji.edu.cn

Abstract. This paper is concerned with the observer design for a class of switched nonlinear systems with both unknown input and measurement noise. In order to deal with the measurement noise, a new descriptor system with augmented state is introduced, and a new Lyapunov-type condition is derived which can not only decouple the unknown input influence but also ensure the asymptotically stability of the observer error system. Subsequently, a sliding mode observer is constructed, and both the state and unknown input are estimated simultaneously. Finally, one example is given to show the effectiveness of the paper.

1. Introduction

The so-called unknown input observer (UIO) is a dynamic system which is able to reconstruct the states of original systems in the presence of unknown inputs[1]-7]. It is wel-known that the external disturbances, actuator faults or system perturbations can all be treated as unknown inputs. In this sense, UIO design techniques are of great significance in control community, especially in the areas of observer-based fault diagnosis [8-10] and observer-based fault tolerant control[11]-13].

The switched system is a typical hybrid system governed by a series of subsystems and a switching signal which determines the switching from one mode to another at every switching point. Indeed, the switched system is very useful, because it can describe a large number of practical systems including flight control systems, network control systems, DC/DC converters and so on[14]. In the past decades, the stability, controllability and observer design of switched systems have been studied extensively[15-21]. However, in the field of switched system UIO designs, there are only a few results can be found in literature[22-26]. For example, in [22] Koenig et al. investigated a class of discrete switched nonlinear system UIO design problems under arbitrary switching signal. In [23], Bejarano et al. proposed a reduced-order UIO design framework for a class of switched systems. However, in such the paper the observers were constructed under the common Lyapunov function strategies, and thus are with some conservativeness. Besides, Yang et al. discussed the observer designs for the switched systems in which the current mode information is unknown[24]. For a class of switched nonlinear systems, Zhang et al. investigated the full-order and reduced-order UIO designs, and meanwhile discussed the relationship of the existence conditions between full-order and reduced-order observers[25]. In [26], for a class of switched descriptor systems, the authors designed a reduced-order observer and the unknown input reconstruction was also discussed.

It should be noted that most of the results associated with the UIO designs of switched systems mentioned above only focused on the systems with unknown inputs in dynamic equation. However,

IOP Conf. Series: Journal of Physics: Conf. Series **1061** (2018) 012013 doi:10.1088/1742-6596/1061/1/012013

seldom did scholars consider the case that output equation is influenced by measurement noise. And this observation motivates our research.

In this paper, the observer design of a class of switched nonlinear system with unknown input and measurement noise is investigated. The major contributions are summarized as: (1) By treating the measurement noise as an extended state, an augmented descriptor system is obtained, and based on such the descriptor system, some useful equivalent conditions are discussed. (2) A full-order sliding mode observer is constructed, by which both the state and the unknown input can be reconstructed.

The rest of the paper is organized as follows. In Section 2, preliminaries including some assumptions and lemmas are presented. The observer design is discussed in Section 3. In Section 4, a simulation is given to verify the effectiveness of the proposed methods.

2. Preliminaries

Consider a class of uncertain switched systems with both actuator and sensor faults

$$\begin{cases} \dot{x}(t) = A_{\sigma(t)}x(t) + B_{\sigma(t)}f(x(t)) + D_{\sigma(t)}\eta(x(t),t) \\ y(t) = C_{\sigma(t)}x(t) + E_{2\sigma(t)}\omega(t) \end{cases} \quad (1)$$

Where $x(t) \in \mathbf{R}^n$, $y(t) \in \mathbf{R}^p$, $\eta(t) \in \mathbf{R}^r$ and $\omega(t) \in \mathbf{R}^w$ are system state, measured output, disturbance and embedded message vectors, respectively. $f(x(t)) \in \mathbf{R}^s$ stands for the nonlinear term of the system. The piecewise constant map $\sigma: \mathbf{R}^+ \to S = \{1, 2, \cdots, N\}$ is a switched signal function defined as $\sigma(t) = i \in S$ when $t_k \le t < t_{k+1}$, $\{t_k\}_{k=0}^{\infty}$ is the sequence of switching points, and each switching point is the time instant at which the system mode is changing from one mode to another (i.e., $\sigma(t_k^-) \ne \sigma(t_k^+)$). And the N is the number of the subsystems. We assume that the ith subsystem is activated in the time interval of $[t_k, t_{k+1})$, that is, we assume that $\sigma(t) = i, \forall t \in [t_k, t_{k+1})$. When the ith subsystem is activated, we simply denote the system matrices as A_i, B_i, C_i, D_i and E_i. Without loss of generality, we assume that matrices B_i, D_i and E_i are all of full column ranks and C_i has full row rank. Moreover, $\operatorname{rank}\begin{bmatrix} B_i & D_i \end{bmatrix} = s + r$

Definition 1: For a switched signal function $\sigma(t)$ any time open interval (t_1, t_2), let $N_{\sigma(t)}(t_1, t_2)$ be the number of discontinuities of $\sigma(t)$ in (t_1, t_2). If there exists two positive constants N_0 and τ_a satisfying

$$N_{\sigma(t)}(t_1, t_2) \le N_0 + \frac{t_2 - t_1}{\tau_a}, t_2 > t_1 \ge 0 \quad (2)$$

Then we say that $\sigma(t)$ has an average dwell time (ADT) τ_a. Specifically, for the half-open time interval $[0, t)$, we use the notation of $N_{\sigma(t)}$ to denote $N_{\sigma(t)}(0, t)$ in short.

Assumption 1: The nonlinear term $f(x)$ is continuous with respective to x and satisfies Lipschitz condition, i.e. there exists a Lipschitz constant $\gamma > 0$, such that $\|f(x) - f(\hat{x})\| \le \gamma \|x - \hat{x}\|$ holds $\forall\ x, \hat{x} \in \mathbf{R}^n$.

Assumption 2: $\operatorname{rank}\begin{bmatrix} sI_n - A_i & 0 & B_i & D_i \\ C_i & E_i & 0 & 0 \end{bmatrix} = n + s + r + w \ \forall s \in \mathbb{C}, \operatorname{Re}(s) \ge 0$.

If we define some augmented matrices and variables

$\bar{x} = \begin{bmatrix} x^T & \omega^T \end{bmatrix}^T$, $T_i = \begin{bmatrix} I_n & 0_{n \times w} \end{bmatrix}$, $\bar{A}_i = \begin{bmatrix} A_i & 0_{n \times w} \end{bmatrix}$, $\bar{C}_i = \begin{bmatrix} C_i & E_i \end{bmatrix}$, then, system (1) can be rewritten as

$$\begin{cases} T\dot{\bar{x}} = \bar{A}_i\bar{x} + \bar{B}_i f(T\bar{x}) + \bar{D}_i\eta \\ y = \bar{C}_i\bar{x} \end{cases} \tag{3}$$

Lemma 1: There are two matrices $G_i \in \mathbf{R}^{(n+w)\times n}$ and $F_i \in \mathbf{R}^{(n+w)\times p}$ such that $G_iT = I_{n+w} + F_i\bar{C}_i$.

Proof: On the one hand, since matrix E_i has full column rank, so $R_i = \begin{bmatrix} T \\ -\bar{C}_i \end{bmatrix} = \begin{bmatrix} I_n & 0_{n\times w} \\ -C_i & -E_i \end{bmatrix} \in \mathbf{R}^{(n+p)\times(n+w)}$ has full column rank and this means that $R_i^T R_i$ is nonsingular. On the other hand, $G_iT = I_{n+w} + F_i\bar{C}_i$ can be rewritten as $X_iR_i = I_{n+w}$, and by verifying it directly, we find that it has solution of $X_i = \Pi_i R_i^T$, where $\Pi_i = (R_iR_i^T)^{-1}$. Because $X_i = \Pi_i R_i^T = \Pi_i \begin{bmatrix} I_n & -C_i^T \\ 0_{w\times n} & -E_i^T \end{bmatrix}$, we have $G_i = \Pi_i \begin{bmatrix} I_n \\ 0_{w\times n} \end{bmatrix}$, $F_i = -\Pi_i \begin{bmatrix} C_i^T \\ E_i^T \end{bmatrix}$.

Lemma 2: $\operatorname{rank}\begin{bmatrix} sI_{n+w} - G_i\bar{A}_i & S_i \\ \bar{C}_i & 0 \end{bmatrix} = n + w + s + r$ holds for all complex number s with $\operatorname{Re}(s) \geq 0$, here we denote $S_i = G_i\Gamma_i \in \mathbf{R}^{(n+w)\times(r+s)}$ and $\Gamma_i = \begin{bmatrix} B_i & D_i \end{bmatrix}$.

Proof: The proof is similar to that of [27], and here we omit it.

Lemma 3: $\operatorname{rank}(\bar{C}_iS_i) = \operatorname{rank}(S_i)$.

Proof: Since

$$\begin{aligned} n + w + s + r &= \operatorname{rank}\begin{bmatrix} T & \Gamma_i \\ \bar{C}_i & 0 \end{bmatrix} = \operatorname{rank}\begin{bmatrix} G_i & -F_i \\ 0 & I_p \end{bmatrix}\begin{bmatrix} T & \Gamma_i \\ \bar{C}_i & 0 \end{bmatrix} = \operatorname{rank}\begin{bmatrix} G_iT - F_i\bar{C}_i & G_i\Gamma_i \\ \bar{C}_i & 0 \end{bmatrix} \\ &= \operatorname{rank}\begin{bmatrix} I_{n+w} & G_i\Gamma_i \\ \bar{C}_i & 0 \end{bmatrix} = \operatorname{rank}\left(\begin{bmatrix} I_{n+w} & G_i\Gamma_i \\ \bar{C}_i & 0 \end{bmatrix}\begin{bmatrix} I_{n+w} & -G_i\Gamma_i \\ 0 & I_p \end{bmatrix}\right) = \operatorname{rank}\begin{bmatrix} I_{n+w} & 0 \\ \bar{C}_i & -\bar{C}_iG_i\Gamma_i \end{bmatrix} \\ &= n + w + \operatorname{rank}(\bar{C}_iG_i\Gamma_i) \end{aligned} \tag{4}$$

So we have $\operatorname{rank}(\bar{C}_iS_i) = s + r$. On the other hand, we have $s + r \geq \operatorname{rank}(S_i) \geq \operatorname{rank}(\bar{C}_iS_i) = s + r$ and this implies $\operatorname{rank}(\bar{C}_iS_i) = \operatorname{rank}(S_i)$.

Lemma 4: [5]Under Assumption 2, for any symmetric positive definite matrix Q_i, the following algebraic matrix equations

$$\begin{cases} (G_i\bar{A}_i - K_i\bar{C}_i)^T P_i + P_i(G_i\bar{A}_i - K_i\bar{C}_i) = -Q_i \\ S_i^T P_i = H_i\bar{C}_i \end{cases} \tag{5}$$

3. Observer Design

Now, for the system (3), we consider

$$\begin{cases} \dot{z} = N_iz + L_iy + M_if(T\hat{\bar{x}}) + \alpha_{i1}(\hat{\bar{x}}, y) + \alpha_{i2}(\hat{\bar{x}}, y) \\ \hat{\bar{x}} = z - F_iy \end{cases} \tag{6}$$

Where N_i, L_i and M_i are constant matrices which will be determined later, and

$$\alpha_{i1}(\hat{\bar{x}}, y) = \lambda_{i1}M_iH_{i1}(y - \bar{C}_i\hat{\bar{x}}) \tag{7}$$

And

$$\alpha_{i2}(\hat{\bar{x}}, y) = \lambda_{i2}\frac{W_iH_{i2}(y - \bar{C}_i\hat{\bar{x}})}{\left\| H_{i2}(y - \bar{C}_i\hat{\bar{x}}) \right\|} \tag{8}$$

IOP Conf. Series: Journal of Physics: Conf. Series **1061** (2018) 012013 doi:10.1088/1742-6596/1061/1/012013

Respectively, where $H_{i1} \in \mathbf{R}^{s\times(n+w)}$ and $H_{i2} \in \mathbf{R}^{r\times(n+w)}$ are matrices consisting of the first s rows and the last r rows of matrix H_i, respectively. λ_{i1} and λ_{i2} are two positive scalars and λ_{i1} is selected to be large enough so that $\lambda_{i2} > \rho$.

Define extended state estimation error as $e = \bar{x} - \hat{\bar{x}} = G_i T\bar{x} - z$, so the dynamic system of the error e is governed by

$$\dot{e} = N_i e + M_i \tilde{f} + W_i \eta - \alpha_{i1}(\hat{\bar{x}}, y) - \alpha_{i2}(\hat{\bar{x}}, y) \tag{9}$$

Where we set $G_i\bar{A}_i - N_i F_i \bar{C}_i - L_i \bar{C}_i - N_i = 0$ and $M_i = G_i B_i$, and denote $\tilde{f} = f(T\bar{x}) - f(T\hat{\bar{x}})$ and $W_i = G_i D_i$.

Theorem 1: Under Assumptions 1-2, if there exists a constant $\mu_1 > 0$, such that $\forall (i, j) \in S \times S$, $i \neq j$,

$$P_i < \mu_1 P_j \tag{10}$$

Then system (6) with (7) and (8) is a robust and adaptive observer of system (3), which is able to estimate the extended state $\bar{x}$ asymptotically, for any switching signal $\sigma(t)$ satisfying $\tau_a \geq \dfrac{\ln \mu_1}{\mu_2}$ where $\mu_2 = \dfrac{1}{2} \max\limits_{i\in s} \left(\dfrac{\lambda_{\min}(Q_i)}{\lambda_{\max}(P_i)} \right)$.

Proof: Consider Lyapunov function candidate $V = e^T P_i e$, on the other hand, because of (10), we know that $\forall (\sigma(t_k^+) = i, \sigma(t_k^-) = j) \in S \times S$, we have $e^T(t_k^+) P_i e(t_k^+) < \mu_1 e^T(t_k^-) P_j e(t_k^-)$, that is

$$V(t_k^+) < \mu_1 V(t_k^-) \tag{11}$$

On the other hand, the derivative of V along with the error dynamic system (9) is

$$\dot{V} = -e^T Q_i e + 2e^T P_i M_i \tilde{f} + 2e^T P_i W_i \eta - 2e^T P_i \alpha_{i1} - 2e^T P_i \alpha_{i2}$$

The second equation of (5) implies that $P_i M_i = (H_{i1}\bar{C}_i)^T$ and $P_i W_i = (H_{i2}\bar{C}_i)^T$, thus we have

$$\begin{aligned} 2e^T P_i M_i \tilde{f} &\leq 2\left|e^T P_i M_i \tilde{f}\right| \leq 2\left\|H_{i1}\bar{C}_i e\right\| \left\|\tilde{f}\right\| \leq 2\gamma \left\|H_{i1}\bar{C}_i e\right\| \|e\| \leq \frac{\gamma^2}{\varepsilon_{fi}} \left\|H_{i1}\bar{C}_i e\right\|^2 + \varepsilon_{fi} \|e\|^2 \\ 2e^T P_i W_i \eta &\leq 2\left|e^T P_i W_i \eta\right| \leq 2\left\|H_{i2}\bar{C}_i e\right\| \|\eta\| \leq 2\rho \left\|H_{i2}\bar{C}_i e\right\| \end{aligned} \tag{12}$$

Now using (5) and (6) give

$$\begin{aligned} 2e^T P_i \alpha_{i1} &= 2\lambda_{i1} e^T P_i M_i H_{i1}(y - \bar{C}_i \hat{\bar{x}}) = 2\lambda_{i1} \left\|H_{i1}\bar{C}_i e\right\|^2 \\ 2e^T P_i \alpha_{i2} &= 2\lambda_{i2} e^T P_i \frac{W_i H_{i2}(y - \bar{C}_i \hat{\bar{x}})}{\left\|H_{i2}(y - \bar{C}_i \hat{\bar{x}})\right\|} = 2\lambda_{i2} \left\|H_{i2}\bar{C}_i e\right\| \end{aligned} \tag{13}$$

So, $\dot{V} \leq -(\lambda_{\min}(Q_i) - \varepsilon_{fi})\|e\|^2 - (2\lambda_{i1} - \dfrac{\gamma^2}{\varepsilon_{fi}})\left\|H_{i1}\bar{C}_i e\right\|^2 - 2(\lambda_{i2} - \rho)\left\|H_{i2}\bar{C}_i e\right\|$. Set $\varepsilon_{fi} = \dfrac{1}{2}\lambda_{\min}(Q_i)$ and $\lambda_{i1} \geq \dfrac{\gamma^2}{\lambda_{\min}(Q_i)}$, we have $\dot{V} \leq -\dfrac{1}{2}\lambda_{\min}(Q_i) e^T e \leq -\dfrac{1}{2}\dfrac{\lambda_{\min}(Q_i)}{\lambda_{\max}(P_i)} V$. That is we obtain $\dot{V} \leq -\mu_2 V$.

Then, in the next, similar to the proof of [14], if $\tau_a \geq \frac{\ln \mu_1}{\mu_2}$ holds, we can conclude that $\lim_{t\to\infty}\|e(t)\| = 0$ and this ends the proof of the theorem.

4. Simulation Example

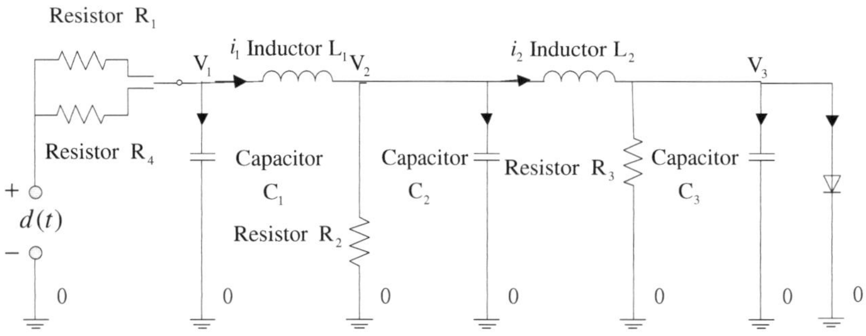

Figure 1 An electronic circuit

In this section, we apply the proposed methods to a switched electronic circuit system. Consider the electronic circuit as is shown in Figure 1, where C_1, C_2 and C_3 stand for capacitors, R_1, R_2, R_3 and R_4 are resistors, L_1 and L_2 are inductors. $d(t)$ is unknown disturbance. The Chua's diode is a nonlinear resistor, whose Voltage-Amperage curve is depicted by

$$I = g(U) = \begin{cases} -1.143U - 0.429, & U \leq -1 \\ -0.714U, & -1 < U \leq 1 \\ -1.143U + 0.429, & U > 1 \end{cases} \tag{14}$$

Where U and I represent the voltage and the current of the Chua's diode, respectively. Then, the circuit system can be described by (1) with

$$A_1 = \begin{bmatrix} 0 & 0 & 1.25 & -1.25 & 0 \\ 0 & 0 & 0 & 1.25 & -1.25 \\ -2.5 & 0 & -2.5 & 0 & 0 \\ 5 & -5 & 0 & -5 & 0 \\ 0 & 1 & 0 & 0 & -0.1 \end{bmatrix}, \ A_2 = \begin{bmatrix} 0 & 0 & 1.25 & -1.25 & 0 \\ 0 & 0 & 0 & 1.25 & -1.25 \\ -2.5 & 0 & -0.125 & 0 & 0 \\ 5 & -5 & 0 & -5 & 0 \\ 0 & 1 & 0 & 0 & -0.1 \end{bmatrix},$$

$$D_1 = \begin{bmatrix} 0 \\ 0 \\ 2.5 \\ 0 \\ 0 \end{bmatrix}, \ D_2 = \begin{bmatrix} 0 \\ 0 \\ 0.125 \\ 0 \\ 0 \end{bmatrix}, B_1 = B_2 = \begin{bmatrix} 0 \\ 0 \\ 0 \\ 0 \\ -1 \end{bmatrix}$$

By choosing $C_1 = 400mF$, $C_2 = 200mF$, $C_3 = 1F$, $R_1 = 1\Omega$, $R_3 = 10\Omega$, $R_4 = 20\Omega$, $L_1 = 800mH$, $L_2 = 800mH$, $\omega(t) = 0.8\sin(t-2)+2$, $\eta(t) = 0.3\cos(0.3t)$. Then, the observer can be designed under the ADT switching signals with $\tau_a \geq \tau_a^* = 2.77$ ($\mu_1 = 2$ and $\mu_2 = 0.25$). Set the initial values $x(0) = [1 \ \ -1 \ \ 0 \ \ -0.5 \ \ -1]^T$ in system (1), and the observer initial value

$z = [0 \ \ 0 \ \ 0 \ \ 0 \ \ 0 \ \ 0]^T$. We plot the observation effectiveness in Figures 2-3. Figure 2 is the switching signal curve. From Fig. 3, we can see that both the state and the measurement noise can be estimated satisfactorily.

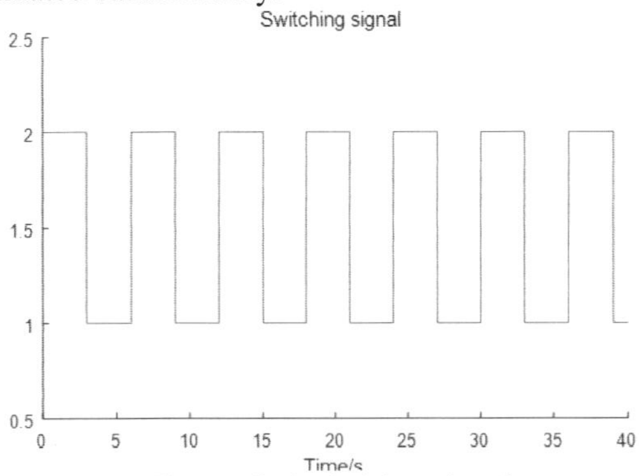

Figure 2. Switching signal

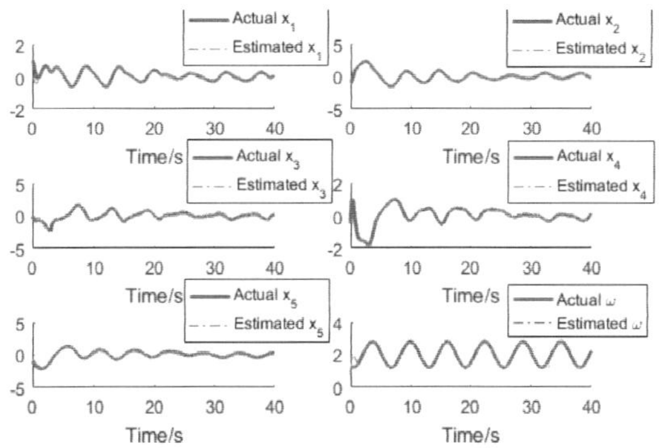

Figure 3. State estimation

5. Conclusion

In this paper, the observer design of a class of switched nonlinear systems with both the unknown inputs and measurement noise are investigated. By considering the measurement noise as a part of state, the original system is transformed into a new descriptor system, and based on such the system some useful conditions which play important roles in observer design are discussed. Following the discussions, a sliding mode observer is constructed by which both the state and the measurement noise are reconstructed. The reduced-order observer design of the switched system with both unknown and measurement noise is an interesting issue, and will be next consideration.

Acknowledgement

This work is supported by the National Natural Science Foundation of China under Grant 61573256 and by the Project of the New Mode of Intelligent Manufacturing.

Reference

[1] Kudva P., Viswanadham N. and Ramakrishna A. Observers for linear systems with unknown inputs, IEEE Trans.Autom. Control, 1980, 25(1): 113-115.
[2] Darouach M., Zasadzinski M. and Xu S. J. Full-order observers for linear systems with unknown inputs, IEEE Trans.Autom. Control, 1994, 39(3): 606-609.
[3] Hou M. and Muller P. C. Disturbance decoupled observer design: A unified viewpoint, IEEE Trans.Autom. Control, 1994, 39(6): 1338-1341.
[4] Hui S. and Zak S. H. Observer design for systems with unknown inputs, Int. J. Appl. Math. Comput. Sci., 2005, 15: 431-446.
[5] M. Corless and J. Tu, State and input estimation for a class of uncertain systems, Automatica, Vol. 34, No. 6, 1998, 757-764.
[6] Zhu F. State estimation and unknown input reconstruction via both reduced-order and high-order sliding mode observers[J]. Journal of Process Control, 2012, 22(1): 296-302.
[7] Zhang J, Zhu F. On the observer matching condition and unknown input observer design based on the system left-invertibility concept[J]. Transactions of the Institute of Measurement and Control, 2017: 0142331217711494.
[8] Alwi H, Edwards C. Robust fault reconstruction for linear parameter varying systems using sliding mode observers[J]. International Journal of Robust and Nonlinear Control, 2014, 24(14): 1947-1968.
[9] Edwards C., Spurgean S. K. and Patton R. J. Sliding mode observers for fault detection and isolation, Automatica, 2000, 36(4): 541-553.

[10] Tan C. and Edwards C. Sliding mode observers for robust detection and reconstruction of actuator and sensor faults, Int. J. Robust Nonlinear Control, 2003, 13(5): 443-463.

[11] Li X., Zhu F., Chakrabarty A. and Zak S. H. Non-fragile fault-tolerant fuzzy observer-based controller design for nonlinear systems. IEEE Transactions on Fuzzy Systems, 2016, 24(6): 1679-1689.

[12] Zhang J, Zhu F. Output control of MIMO system with unmatched disturbance based on high-order unknown input observer[J]. International Journal of Control, Automation and Systems, 2017, 15(2): 575-584.

[13] Zhang J, Zhu F. Observer-based output consensus of a class of heterogeneous multi-agent systems with unmatched disturbances[J]. Communications in Nonlinear Science and Numerical Simulation, 2018, 56: 240-251.

[14] Zhao X, Zhang L, Shi P, et al. Stability and stabilization of switched linear systems with mode-dependent average dwell time[J]. IEEE Transactions on Automatic Control, 2012, 57(7): 1809-1815.

[15] Lin H, Antsaklis P J. Stability and stabilizability of switched linear systems: a survey of recent results[J]. IEEE Transactions on Automatic control, 2009, 54(2): 308-322.

[16] Daafouz J, Riedinger P, Iung C. Stability analysis and control synthesis for switched systems: a switched Lyapunov function approach[J]. IEEE transactions on automatic control, 2002, 47(11): 1883-1887.

[17] Zhao X, Zheng X, Niu B, et al. Adaptive tracking control for a class of uncertain switched nonlinear systems[J]. Automatica, 2015, 52: 185-191.

[18] Niu B, Zhao J. Barrier Lyapunov functions for the output tracking control of constrained nonlinear switched systems[J]. Systems & Control Letters, 2013, 62(10): 963-971.

[19] Zhao X, Yu Z, Yang X, et al. Estimator design of discrete-time switched positive linear systems with average dwell time[J]. Journal of the Franklin Institute, 2014, 351(1): 579-588.

[20] Tanwani A, Shim H, Liberzon D. Observability for switched linear systems: characterization and observer design[J]. IEEE Transactions on Automatic Control, 2013, 58(4): 891-904.

[21] Zhao X, Liu H, Zhang J, et al. Multiple-mode observer design for a class of switched linear systems[J]. IEEE Transactions on Automation Science and Engineering, 2015, 12(1): 272-280.

[22] Koenig D, Marx B, Jacquet D. Unknown input observers for switched nonlinear discrete time descriptor systems[J]. IEEE Transactions on Automatic Control, 2008, 53(1): 373-379.

[23] Bejarano F J, Pisano A. Switched observers for switched linear systems with unknown inputs[J]. IEEE Transactions on Automatic Control, 2011, 56(3): 681-686.

[24] Yang J, Chen Y, Wang X. Active mode identification and continuous state estimation for switched linear systems with unknown inputs and slow switching signal[J]. Circuits, Systems, and Signal Processing, 2015, 34(7): 2193-2211.

[25] Zhang J, Zhu F, Zhao X, et al. Robust impulsive reset observers of a class of switched nonlinear systems with unknown inputs[J]. Journal of the Franklin Institute, 2017, 354(7): 2924-2943.

[26] Hou Y, Zhu F, Zhao X, et al. Observer Design and Unknown Input Reconstruction for a Class of Switched Descriptor Systems[J]. IEEE Transactions on Systems, Man, and Cybernetics: Systems, 2017.

[27] Zhang J.; Zhu F., Linear system unknown input observer design when the observer matching condition is not satisfied[J], Control Theory & Applications , 2017, 34(4) 441-448.

IOP Conf. Series: Journal of Physics: Conf. Series **1061** (2018) 012014 doi:10.1088/1742-6596/1061/1/012014

Optimization of Deep Convolution Neural Network Based on Sparse DropConnect

Liu Mengxi[1, 2, *], Song Jiuxu[2], Wang Zheng[2] and Ju Yongfeng[1]

[1]School of Electronic and Control Engineering, Chang'an University, Xi'an, 710064, China;

[2] Shaanxi Key Laboratory of Measurement and Control Technology for Oil and Gas Wells,Xi'an,710065,China

*E-mail:26702163@qq.com

Abstract: According to the problem of overfitting in the traditional Convolutional Neural Networks(CNN) with large samples, an improved sparse DropConnect algorithm is put forward. With proposed algorithm, the CNN is improved and optimized. The performance of the network in image recognition is verified. Experiments of weld image recognition and classification indicate that the selective ability of the CNN on the sparse feature is improved. High recognition accuracy and classification accuracy of test sample are achieved. Over-fitting in the traditional network model can be avoided. More important, the feature sparsity and discriminative ability of the CNN are increased.

1. Introdunction

Deep Convolutional Neural Networks (CNN), as one of the deep learning models, is suitable for processing of large-scale image data. With regard to CNN feature extraction ability, rapid increase and decrease can be achieved by adjustment of the number of network layers and the number of neurons per layer [1]. In terms of application, when passing each layer, CNN input data will obtain corresponding significant features of observation data, which include significant features such as translation, scaling, and rotation invariance. Then, by means of layer-by-layer extraction, advanced features of the image can be extracted at a high level and successfully applied to target detection, image classification, speech recognition and other fields[2]. At present, many studies on application scenarios for image recognition and classification are centered on the optimization of deep network structure. there are improved RCNN [3], Faster-RCNN [4], and YOLO [5] network structures that break the limitations of traditional Convolutional Neural Networks through area selection.

However, large image data sets have a great influence on the images recognition speed, and the network scale of CNN also increases, making the abstract relationship between training samples and features more and more complex. The BP algorithm enables the CNN to start from a randomly initialized starting point and achieves a near-perfect recognition result on a limited training sample after a sufficient number of iterations. Due the feature extractor is trained too well for the training sample, there are deviations from the image features in the testing sample, resulting in overfitting [6]. This requires CNN to be optimized, considering enhancement of the network performance by means of improving the sparsity of the CNN.

2. Optimization design of sparse CNN

In order to prevent over-fitting between training sample and test sample, this paper introduces DropConnect [7] algorithm, which schematic diagram as shown in the figure 1.In the DropConnect algorithm, the connection weights between the two networks are zeroed by a certain probability, and the neuron weight mask discard is generated by a Bernoulli variable, which is independent of the neuron weights and activation values. Compared with the DropOut algorithm, the DropConnect algorithm has better accuracy and efficiency in image recognition and classification.

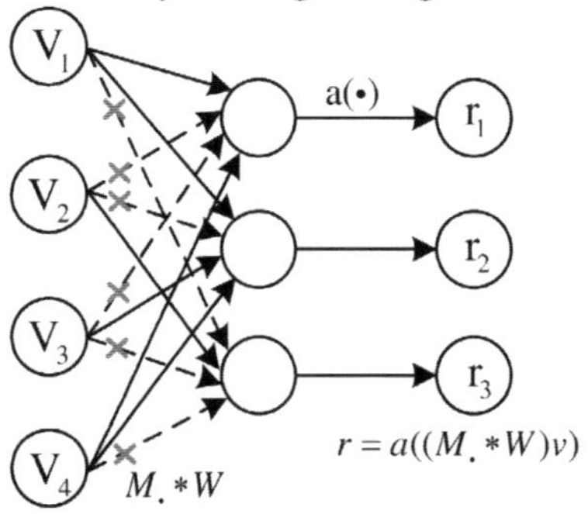

Figure 1 . DropConnect structure diagram

2.1 DropConnect sparsity investigation

To introduce a sparse mechanism in the DropConnect algorithm, all weights are dropped with the same probability. If different neurons allocate different drop probabilities according to the sparsity of the activation value, the weight mask becomes the function of the activation value of the neuron. The improvement enables the network model to dynamically adjust the drop probability of the weight value during the training process according to the sparsity of the output value of the network layer, and thus dynamically adjust the statistical model of the network, allowing the neuron connection with sparse feature extraction capability to have greater probability of being activated, eventually leading more sparse features of network model extraction [8]. The Sparse DropConnect training algorithm can be described as :

$$\text{mask}_{ij}^{(1)} \sim \text{Uniform}(0.5) \tag{1}$$

$$\rho_{ij}^{(l)} = p_{drop}(y_j^{(l-1)}) \tag{2}$$

$$\tilde{W}_{ij}^{(l)} = \tilde{W}_{ij}^{(l)} \circ (mask_{ij}^{(l)} > \rho_{ij}^{(l)}) \tag{3}$$

$$z_i^l = \sum_{j=1}^{n} \tilde{W}_{ij}^{(l)} y_j^{(l-1)} + b_i^{(l)} \tag{4}$$

$$y^{(l)} = f(z^l) \tag{5}$$

$$\sum_M f((M \circ W)x) \approx f(\sum_M (M \circ W)x) \tag{6}$$

Where, $W_{ij}^{(l)}$ is for the connection weight between the j neuron in the l-1 layer and the i neuron in the l layer; $P_{ij}^{(l)}$ is for the drop probability of the weight associated with the weight $W_{ij}^{(l)}$, and the more sparse neuron connection has a lower drop probability; $P_{drop}(\cdot)$ is for the calculation function of weight drop probability; The sparsity measurement method used in this paper is based on the relationship between L_1 norm and L_2 norm [9]. The specific formula is:

$$sparseness(x) = \frac{\sqrt{n} - (\sum |x_i|) / \sqrt{\sum x_i^2}}{\sqrt{n} - 1} \tag{7}$$

Where, x is the input feature vector, n is the number of elements in the vector x. After normalization processing, the output value of the sparsity formula is in the range [0, 1.0].

2.2 CNN optimization.

The deep Convolutional Neural Network with improved structure is shown in Figure 2.

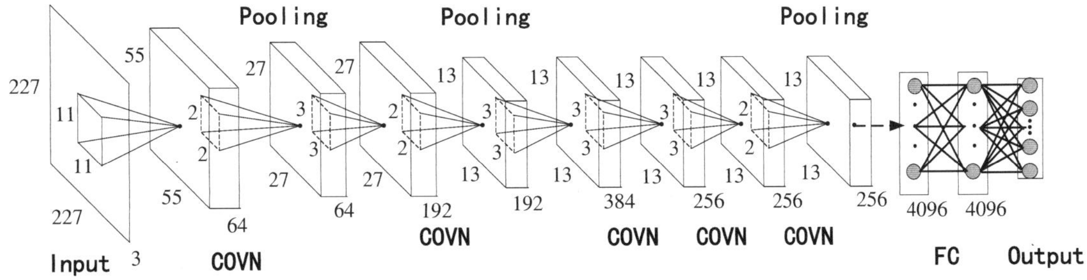

Figure 2. Structure diagram of deep Convolutional Neural Network

In the network, the settings of COVN layer, Pooling layer, and FC layer are shown in Table 1. The last layer uses the Softmax function to output the probability of the category that each dimension image belongs to. After addition of Sparse DropConnect to the f FC layer, the initial learning rate is 0.001, the momentum is 0.9, the weight attenuation is 0.0005, and the number of iterations is 500 [10].

Table 1. Parameter settings of CNN

Layer No.	COVN layer	Pooling layer	FC layer
1	64×11×11 stride1 pad0	2×2 maxpool	4096 softmax:5
2	192×3×3 stride2 pad2	2×2 maxpool	4096 softmax:2
3	384×3×3 stride1 pad1		
4	256×3×3 stride1 pad0		
5	256×3×3 stride1 pad1	2×2 maxpool	

3. Numerical experiments

To verify the algorithm proposed in this paper, X-ray image data of steel pipe welds taken in industrial production are selected for the test to effectively identify and classify weld defects images. The size of the image is randomly selected. To maximum retention the characteristic information ,when the image is processed, the longer side of the length and width is converted to 256 pixels, and the other side is subjected to the equal proportion conversion [11], which results in the other side being less than 256 pixels. For the shorter side, the original image is centered, and both sides are overturned to a size of 256 pixels. On the one hand, the methord ensures that the processed image is still 256 × 256 in size, and on the other hand, it will not cause the loss of welds defects information in the image. The specific transformation algorithm is shown below:

Input:InImage

Output:OutImage

Step1:Compute the image length and width ,the longer side denoted as:ma, the shorter side denoted as mi;

Step2:If ma>256, go to the equal proportion conversion by 256/ma, then step4; or Step3:

Step3:Overturn ma, until 256;

Step4:Overturn mi, until 256;

Step5:Output outImage

The experiment uses the CNN structures and parameters proposed in Section 2.2, and uses Softmax to connect and implement image classification. To better verify the fitting, the ratio of the training sample to the testing sample in the original defect sample data is set to 1:1, 500 samples respectively.

To study the influence of sparsity on CNN, Sparse DropConnect and DropConnect after the FC layer of CNN were respectively used to perform image recognition performance experiments. The results are shown in the figure 3. The figure shows the experimental results of the variation of recognition performance with sparsity. From the figure 3, it can be seen that with the change to sparsity, CNN, to which DropConnect was added, experience large fluctuations in recognition performance, while Sparse DropConnect is less affected by sparsity. This is because sparse action in Sparse DropConnect is also in fact a kind of regularization operation. It has a similar effect in preventing model over-matching and therefore has less fluctuation.

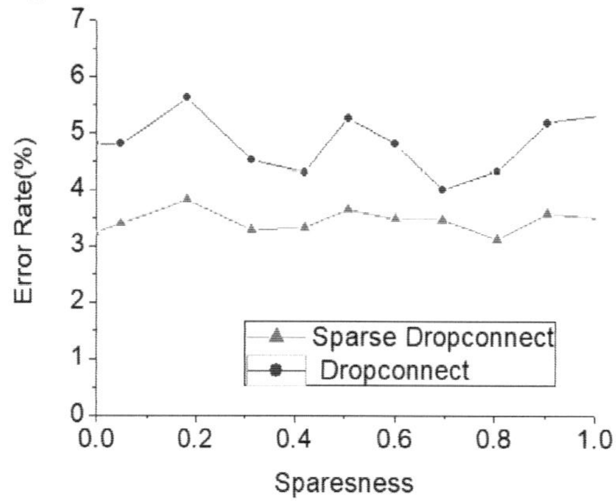

Figure 3. Relationship between sparsity and error rate

In the image recognition fitting experiment, comparison experiments were performed using the Sparse DropConnect, DropConnect, DropOut, and non- regularization measures after the FC layer of CNN. The experimental results are shown in Figure 4. As seen from Figure 4, models that did not use the regularization method soon experienced overfitting during the training process. While the models using the Sparse DropConnect, DropConnect, and DropOut methods slowed down, they eventually achieved better recognition results. The rate of convergence of Sparse DropConnect model is slower than DropOut, slightly faster than DropConnect, and has the lowest error rate of the test set, avoiding overfitting very well.

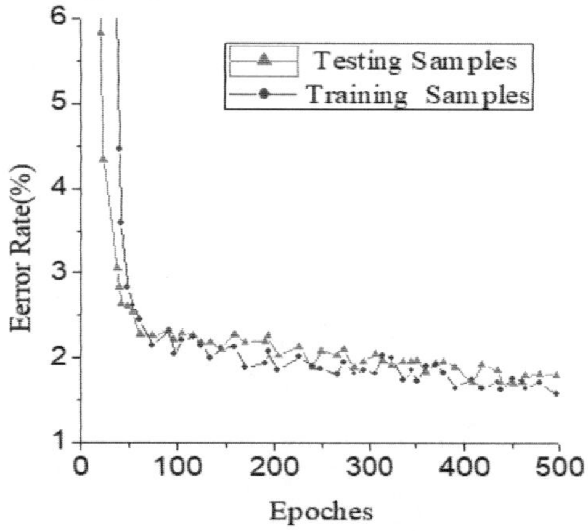

(1) Sparse DropConnect

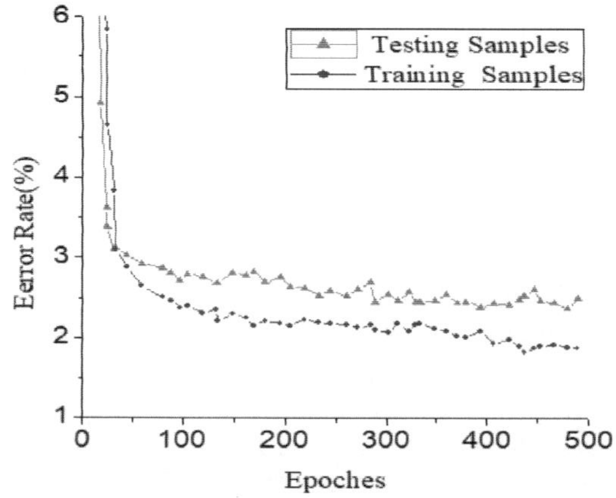

(2) DropConnect

IOP Conf. Series: Journal of Physics: Conf. Series **1061** (2018) 012014 doi:10.1088/1742-6596/1061/1/012014

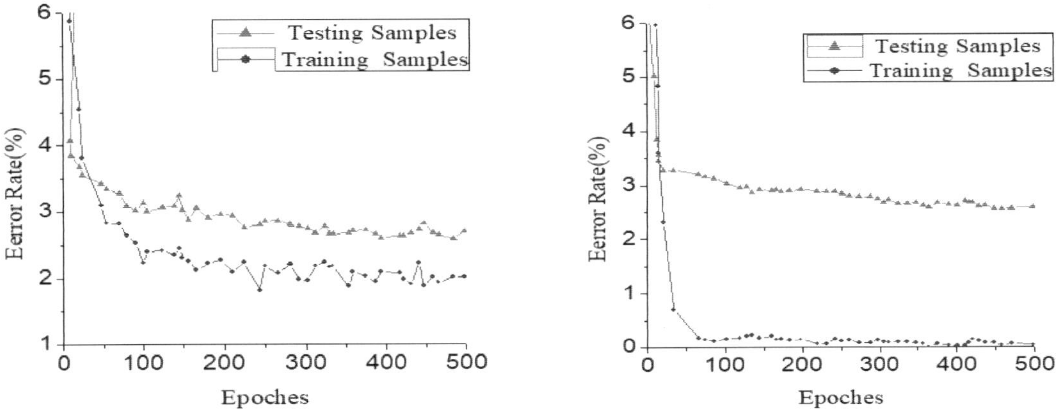

(3) DropOut (4) No rules

Figure 4. CNN training and testing error rate

The main types of welds defects included: porosity (1), cracks(2), slag inclusions(3), and non-fusion (4), incomplete penetration(5). In the classification and recognition experiment for weld defects, comparison experiments were performed using the Sparse DropConnect, DropConnect, DropOut, and non- regularization measures respectively. The experimental results are shown in Table 2, the results show that: CNN under Sparse DropConnect has certain advantages, mainly the accuracy of the two training and test samples higher than that of other methods and the data of two samples tending to be consistent, showing a better goodness of fit.

Table 2. Weld defects classification results

Defect type	Testing sample recognition rate(%)				Training sample recognition rate(%)			
	No-Drop	Dropout	Drop Connect	Sparse DropConnect	No-Drop	Dropout	Drop Connect	Sparse DropConnect
1	**89.45**	**92.36**	**91.43**	**96.77**	**88.45**	**91.45**	**92.22**	**96.98**
2	**89.44**	**92.45**	**92.16**	**95.26**	**90.44**	**93.89**	**95.29**	**95.86**
3	**90.78**	**92.78**	**91.78**	**95.73**	**93.29**	**94.46**	**89.19**	**96.03**
4	**90.86**	**91.89**	**89.86**	**93.00**	**92.57**	**95.19**	**92.49**	**93.16**
5	**90.28**	**91.59**	**92.36**	**92.86**	**91.45**	**95.28**	**95.22**	**93.10**

Compared with other methods, the CNN network under Sparse DropConnect in the test has higher recognition and classification of the weld defects images. The first reason is that the deep network structure of CNN determines that the transmission of data information is made on the bottom-to-top and layer-by-layer basis, retaining important information as much as possible and representing it in multiple levels, so it has a better recognition accuracy than the shallow network; secondly, the introduction of sparse DropConnect in the CNN network can obtain sparse representation of the target, reduce information processing amount, and simplify the learning and training process, greatly improving the accuracy and adaptability of defect feature classification.

4. Conclusion

As a kind of deep learning method, CNN has its better image feature extraction and information expression capability. It expands and optimizes the the CNN structure by sparse DropConnect algorithm added after the FC, increasing the ability of network to select sparse features. The optimized CNN ensures higher recognition accuracy under the precondition of increase the sparsity of the output features. As the data set of the weld image collected by the industry used in this experiment is not big enough, further experimental studies are needed for more data sets under the CNN.

IOP Conf. Series: Journal of Physics: Conf. Series **1061** (2018) 012014 doi:10.1088/1742-6596/1061/1/012014

References

[1] Simard P Y , D Steinkraus, and J C Platt 2003 Best practices for convolutional neural networks applied to visual document analysis *Proceedings* IEEE *International Conference on Document Analysis and Recognition* 958

[2] Soatto S and Chiuso A 2016 Visual representations: defining properties and deep approximations *Eprint Arxiv* **2018** 161-9

[3] Ioffe S and Szegedy C 2015 Batch normalization: accelerating deep network training by reducing internal covariate shift *International Conference on International Conference on Machine Learning* 448-56

[4] Ren S,He K, Girshick R and Sun J 2017 Faster r-cnn: towards real-time object detection with region proposal networks. *IEEE Trans Pattern Anal Mach Intell* ***39*(6)** 1137-49.

[5] Redmon J, Divvala S, Girshick R and Farhadi A 2016 You Only Look Once: Unified, Real-Time Object Detection. *IEEE Conference on Computer Vision and Pattern Recognition* 779-88

[6] Gong Y, Wang L, Guo R and Lazebnik S 2014 Multi-scale orderless pooling of deep convolutional activation features *European Conference on Computer Vision* **8695** 392-407.

[7] Pennec X 2008 Statistical computing on manifolds: from riemannian geometry to computational anatomy. *Lecture Notes in Computer Science* ***5416*** 347-86.

[8] Goodfellow I J, Bulatov Y, Ibarz J, Arnoud S and Shet V 2013 Multi-digit number recognition from street view imagery using deep convolutional neural networks *Computer Science* 29

[9] Zha S, Luisier F, Andrews W, Srivastava N and Salakhutdinov R 2015 Exploiting image-trained cnn architectures for unconstrained video classification *Computer Science* 106

[10] Zhang N, Paluri M, Ranzato M, Darrell T and Bourdev L 2014 PANDA: Pose Aligned Networks for Deep Attribute Modeling. *Computer Vision and Pattern Recognition* 1637-44

[11] Mendel K R,Li H, Sheth D and Giger M L 2018 Transfer learning with convolutional neural networks for lesion classification on clinical breast tomosynthesis *Computer-Aided Diagnosis* 28

IOP Conf. Series: Journal of Physics: Conf. Series **1061** (2018) 012015 doi:10.1088/1742-6596/1061/1/012015

Aircraft Type Recognition Based on Target Track

Honglan Huang, Jincai Huang, Yanghe Feng, Zhong Liu, Tao Wang, Li Chen and Yuzhen Zhou

College of Systems Engineering, National University of Defense Technology, Changsha, Hunan, PR China

fengyanghe@nudt.edu.cn

Abstract. The target recognition plays an important role in air traffic management, and the research on automatic aircraft recognition is still in the exploratory stage. Since aircrafts move at high speeds in complex background, it is still a challenging task to quickly extract valid features from small amounts of data for aircraft type recognition. The machine learning is capable of capturing discriminating information and can well identify potential patterns from the dataset. Thus, we employ a novel classification model based on machine learning with the use of some effective motion features extracted from aircraft flight track information as the input of the model such as maximum speed, cruise speed, maximum acceleration, maximum climb rate. We performed experiments on datasets collected by multiple sensors and the results demonstrated the effectiveness of proposed model in significantly improving the aircraft type recognition accuracy, especially for military aircrafts. Besides, the model avoids the inherent deficiencies of graphics and image processing, and can well meet the needs of the air traffic management.

1. Introduction

With the increasing types and numbers of aircrafts, the ability to reliably recognize aircraft type is a vital aspect of air traffic control in both daily life and modern battlefield. At present, in order to achieve a better recognition accuracy, the work of aircraft type recognition still needs a great deal of human experience and consumes a lot of time and other manpower resources. A common technique for identification of military aircraft is Identification Friend Foe (IFF). Civilian aircraft use a technique similar to IFF called Secondary Surveillance Radar (SSR)[1]. The fundamental drawback of techniques like IFF and SSR is that they require active cooperation of pilots which makes these techniques less efficient and less practical. The research on automatic aircraft recognition is still in the exploratory stage, and most of the existing work is based on graphic image processing[2-5]. Aircraft recognition based on image contour is mainly to find the approximate invariant features[6-8]. Commonly used invariant feature extraction methods include Hu matrix[7], affine distance, Fourier descriptor[9], wavelet moment, and Zernike moments. Radar signals are also widely used in air traffic control[1]. Both radar-based and image-based methods make use of the distinctions of aircraft contour to identify the type of the planes. However, in practice, these methods face many challenges: 1. The overall shapes of the aircraft are generally similar, especially for a wide range of civilian aircrafts. 2. The aircraft's shape is highly dependent on the orientation and distance of the aircraft with respect to the sensor. 3. It is difficult to take a clear picture of an aircraft moving in high speed. 4. Image quality is heavily constrained by the weather and other natural factors. These challenges make the identification method based on contour information more difficult to achieve in practical application.

To overcome these serious drawbacks, we employ a novel classification model based on machine learning with the use of some effective motion features, such as maximum speed, cruise speed, maximum acceleration, maximum climb rate, extracted from aircraft flight track information as the input of the model. Different from extracting contour invariants during aircraft movement, we extract dynamic characteristics during the movement which can well avoid the shortcomings mentioned before. This model has several obvious advantages: 1. Motion features have strong representation, and identifying aircraft type by its set characters on movement is closer to the real needs. 2. It is easier to extract dynamic motion features than invariant features in a dynamic process[10], and the model of a single aircraft's movement can be reduced to a movement model of a probe point which can greatly reduce the computational complexity, speed up the recognition process and increase the ability to quickly handle the task of identifying a large number of targets. 3.Since most aircrafts fly on fixed routes, by combining with the track information, our model makes good use of the historical experience and knowledge, and can obtain higher accuracy on civilian aircrafts whose shape and movement characteristics are very similar.

2. Basic approach

2.1. Data collection and preprocessing

Our datasets are based on the time-series data of aircraft flight track collected by multiple sensors including the detecting time *t*, longitude *Lng*, latitude *Lat*, altitude *H*, velocity *V*, heading *Di*, sensor signal type *Sg* and aircraft type label *Label*. And the aircraft type labels are manually marked according to expert's experience. Here we think there is no error in the marked labels.

2.2. Feature extraction

Due to the differences in aircraft motion performance and pilot's flying habit, we can extract useful features as the input of the recognition model based on historical track. Wang M[11] discussed nine motion performance features that may affect the aircraft type recognition, but some of them are complicated to calculate and require high precision of the acquisition sensor. Therefore, in this paper, four motion features were extracted from the flight data: maximum speed, cruising speed, maximum acceleration, maximum rate of climb.

Maximum speed: Aircraft reached the maximum speed when the booster was at maximum thrust. At the maximum speed, the tail of the airplane cannot be heat balanced and this state cannot last long. In our data set, we will approximate the detected instantaneous maximum speed as the maximum speed of the aircraft.

$$V_k max = max_i \, v_{ki} \tag{1}$$

Where *k* is the number of the aircraft in the detection record and *i* is the ith detection time point.

Cruising speed: Also called as economic speed, when the aircraft did not open afterburner, the aircraft can stay in the air the longest time in this state and the tail of the airplane can reach thermal equilibrium. We approximate the average speed of the probe to the cruising speed of the aircraft.

$$\overline{V_k} = \frac{\sum_{i=1}^{n} v_{ki}(t_{k(i+1)} - t_{ki})}{t_{kn} - t_{k1}} \tag{2}$$

Where *n* is the number of probe records of aircraft *k*.

Maximum acceleration: The maximum capacity of the aircraft to enhance speed. The magnitude of the acceleration is related to the propeller's power, shape and cooling capacity of aircraft. Therefore, the ability of acceleration is an important parameter that reflects the characteristics of different types. We calculate the difference of velocity in each probe interval as the value of acceleration and choose the maximum value as the maximum acceleration characteristic of the aircraft.

$$a_{kv} max = \max_i \frac{v_{k(i+1)} - v_{ki}}{t_{k(i+1)} - t_{ki}} \tag{3}$$

Maximum rate of climb: The maximum rate of climb reflects the ability of the aircraft to overcome its gravity and resistance and is one of the characteristics that can best reflect the kinematic performance of the attack aircraft. In the flight track information, the maximum value of altitude difference in the detection interval is considered as the maximum rate of climb (RoC).

$$\mathrm{RoC}max = \max_i \frac{h_{k(i+1)}-h_{ki}}{t_{k(i+1)}-t_{ki}} \tag{4}$$

Since most aircrafts have a fixed route, we consider the longitude, latitude, altitude, velocity, heading information, sensor signal types together with the above four characteristics as the inputs of our recognition model.

2.3. Aircraft type recognition model using machine learning

The architecture of the proposed recognition model is demonstrated in Figure 1. The input of the model are the selected features. The classic algorithms in machine learning including SVM[12, 13], ELM[14-16] and RKELM[17] are used respectively in our model as the classifier. The output of the model is the type of aircraft.

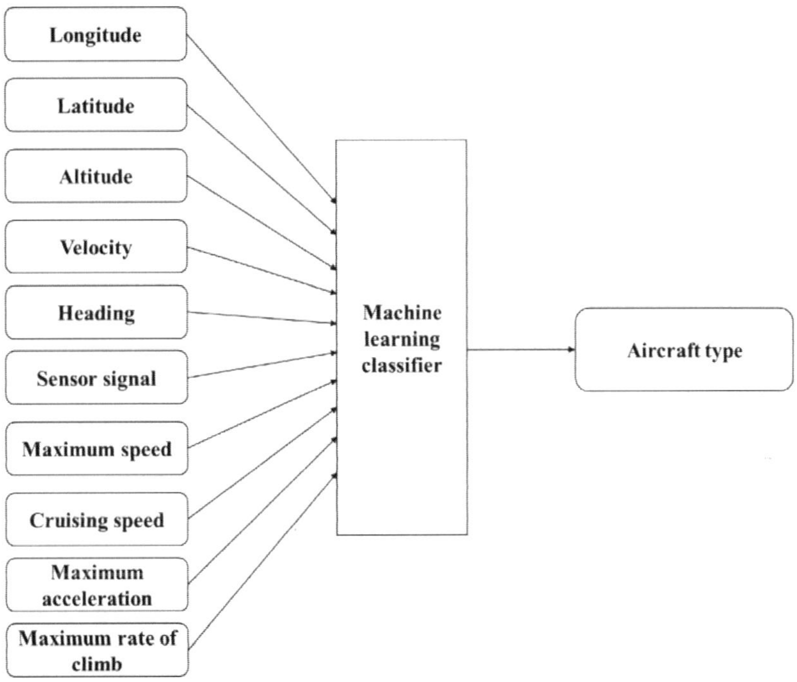

Figure 1. Proposed aircraft recognition model

3. Experiments and results

3.1. Datasets

We conducted experiments on four different datasets respectively: dataset 1 covers 3 classes of 26309 samples; dataset 2 covers 4 classes of 10928 samples; dataset 3 covers 7 classes of 3020 samples; dataset 4 covers 10 classes of 3910 samples. The dataset 1 is collected from civilian aircrafts of the Boeing Company, Airbus and Bombardier respectively. The dataset 2 is also collected from civilian aircrafts. The dataset 3 is collected from military aircrafts, while data set 4 contains both track information of military and civilian aircrafts. As part of the data is classified, we use numbers to indicate the types of aircraft. Figure 2 shows the category distribution of these four datasets.

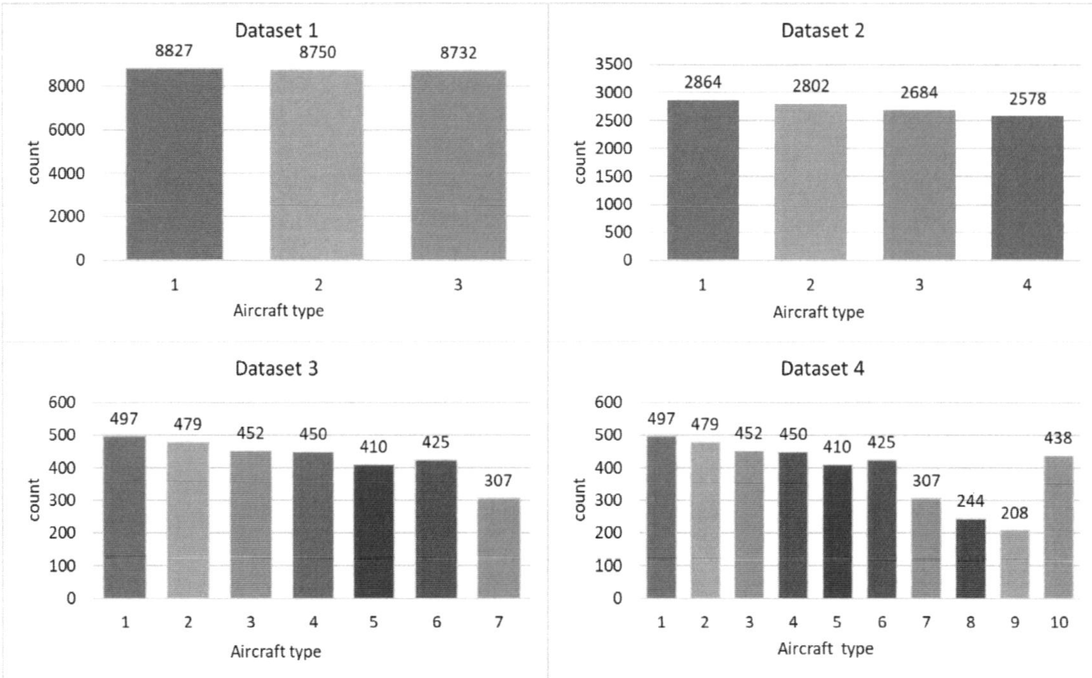

Figure 2. Category distribution of datasets

3.2. Methodology

All datasets are divided into training set (70%) and test set (30%), and all the datasets are processed in two ways. One way does not extract the motion characteristics from the flight data, which takes track information directly as the input of the model. The other way adds the extracted motion features to the original data, and then the recognition model takes this data as input. The coefficient correlation matrix showed the correlation among all the features and the aircraft type. As can be seen in Figure 3, the relative high correlation between the movement characteristics we extracted and the aircraft type indicates that the movement characteristics make a positive contribution to the aircraft type recognition. The significant correlation between the longitude and the type of sensor signal is due to the difference in geographic location of the sensors.

For the classical ELM algorithm, we use the Hardlim function as the activation function.

$$\text{hardlim}(x) = \begin{cases} 1, & if\ x \geq 0 \\ 0, & otherwise \end{cases} \tag{5}$$

For RKELM algorithm, the Gaussian kernel function has been used in the experimental study.

$$k(\mathbf{x}, \mathbf{x}_i) = \frac{\exp\left(-||\mathbf{x}-\mathbf{x}_i||^2\right)}{\sigma} \tag{6}$$

For SVM algorithm, Gaussian kernel function has also been considered in experiment.

$$k(\mathbf{x}, \mathbf{x}_i) = \exp\left(-\gamma||\mathbf{x} - \mathbf{x}_i||^2\right) \tag{7}$$

In addition, all the results reported are the average of 20 independent trials. For ELM and RKELM networks, the number of hidden nodes l is chosen by cross-validation method with step size 10, and only the obtained optimized l is reported. In order to achieve better generalization performance, the parameters of the model need to be properly selected. After a large number of experiments, we choose the parameter values shown in Table 1.

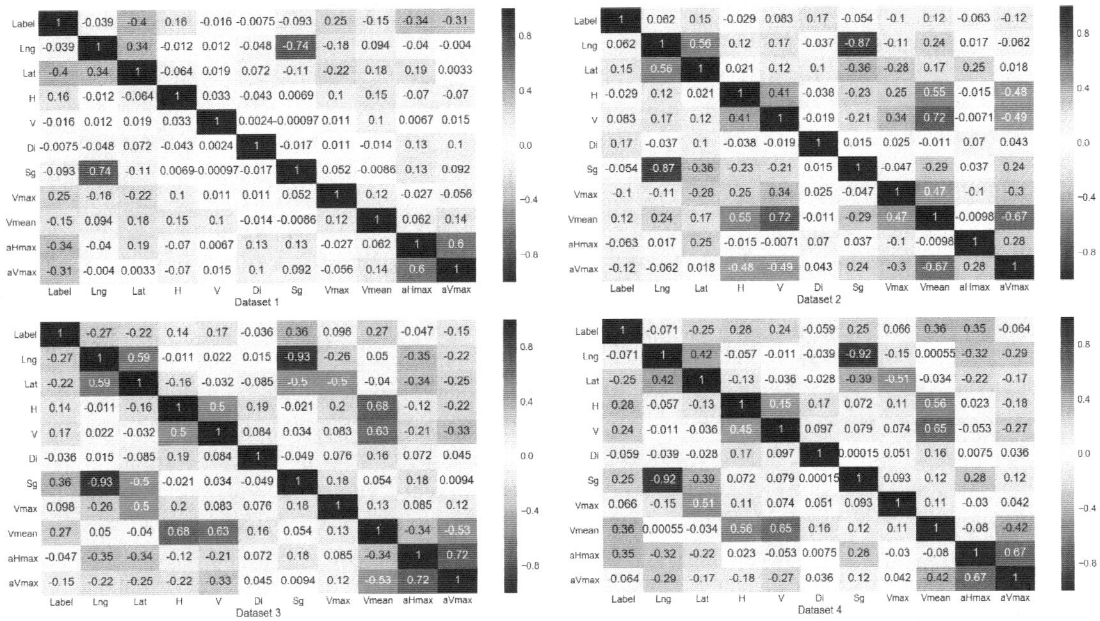

Figure 3. Correlation coefficient matrix thermogram of datasets

Table 1. Parameters on datasets

Dataset	Classes	Number of samples	ELM	RKELM	SVM
			l	(l, C, σ)	(C, γ)
1	3	26309	3000	$(2000, 2^{20}, 2^{20})$	(1000,0.00001)
2	4	10928	3000	$(400, 2^{22}, 2^{22})$	(100,0.00001)
3	7	3020	3000	$(400, 2^{24}, 2^{24})$	(100,0.00001)
4	10	3910	3000	$(400, 2^{24}, 2^{24})$	(1000,0.00001)

3.3. Performance evaluation

Because there is no obvious category imbalance in our dataset, the accuracy on the test sets was employed to measure the performance of the model. Some interesting remarks can be drawn from Table 2. Firstly, compared with the group without the extracted features, it is possible to observe that the testing accuracy has been significantly improved by adding the extracted motion features. In dataset 1, the testing accuracy of ELM was improved by 22.21%, RKELM outperformed with 17.74% improvement, SVM outperformed with 14.77% improvement. In dataset 2, the testing accuracy of ELM was improved by 12.69%, RKELM outperformed with 11.40% improvement, SVM outperformed with 10.85% improvement. In dataset 3, the testing accuracy of ELM was improved by 15.38%, RKELM outperformed with 6.92% improvement, SVM outperformed with 9.19% improvement. In dataset 4, the testing accuracy of ELM is improved by 16.63%, RKELM outperformed with 11.32% improvement, SVM outperformed with 9.56% improvement. Meanwhile, the training time did not increase significantly. Secondly, the model using SVM algorithm has a significantly better recognition effect on all data sets than the other two algorithms. Since our data volume is small, the disadvantages of SVM in training time are not obvious. When facing a larger training data set, RKELM algorithm can achieve faster training speed and high prediction accuracy. The experimental results are shown in Table 2.

In the meantime, it was found that our model gives better performance for the recognition of military aircraft. This means that our model may have a broad application prospect in the future informationized battlefield.

Table 2. Average recognition accuracy on datasets

Dataset \ Accuracy	No motion features added			Motion features added		
	ELM	RKELM	SVM	ELM	RKELM	SVM
1	60.60%	74.59%	81.58%	**82.81%**	**92.34%**	**96.35%**
2	69.71%	81.79%	87.77%	**82.41%**	**93.19%**	**98.63%**
3	74.78%	88.75%	89.82%	**90.16%**	**95.67%**	**99.02%**
4	72.23%	85.21%	89.58%	**88.86%**	**96.53%**	**99.14%**

4. Conclusion

As an important technology in air traffic management, aircraft type recognition is gaining more and more attention by scholars. The existing researches have been mostly based on graphic image processing, which is inherently deficient in the high dynamic real-time air combat. We employ a novel classification model based on machine learning with the use of some effective motion features extracted from aircraft flight track information as the input that automatically identifies aircraft type. Experimental results show that the model has a good recognition effect, especially for military aircraft. The model avoids the inherent deficiencies of graphics and image processing, so it will well meet the needs of the air traffic management and has a wide range of application in modern battlefield.

Acknowledgments
This study was supported by the National Natural Science Foundation of China (No. 71701205 and No. 71701206).

References

[1] Proefschrift A, Zwart J P, Gelsema I S J, et al. Aircraft Recognition from Features Extracted from Measured and Simulated Radar Range Profiles [J]. Universiteit Van Amsterdam Uva, 2003,
[2] Somaie A A, Badr A, Salah T. Aircraft recognition system using eigenvector technique; proceedings of the Radio Science Conference, 1999 NRSC '99 Proceedings of the Sixteenth National, F, 1999 [C].
[3] Liu Y. Aircraft type recognition based on convex hull features and SVM [J]. Proceedings of SPIE - The International Society for Optical Engineering, 2007, 6786(
[4] Hsieh J W, Chen J M, Chuang C H, et al. Aircraft type recognition in satellite images [J]. IEE Proceedings - Vision, Image and Signal Processing, 2005, 152(3): 307-15.
[5] Kamgar-Parsi B, Kamgar-Parsi B, Jain A K. Automatic aircraft recognition: toward using human similarity measure in a recognition system; proceedings of the Computer Vision and Pattern Recognition, 1999 IEEE Computer Society Conference on, F, 1999 [C].
[6] Dudani S A, Breeding K J, Mcghee R B. Aircraft Identification by Moment Invariants [J]. IEEE Transactions on Computers, 1977, C-26(1): 39-46.
[7] Rong H J, Jia Y X, Zhao G S. Aircraft recognition using modular extreme learning machine [J]. Neurocomputing, 2014, 128(5): 166-74.
[8] Li X D, Pan J D, Dezert J. Automatic Aircraft Recognition using DSmT and HMM; proceedings of the International Conference on Information Fusion, F, 2014 [C].
[9] Wallace T P, Wintz P A. An efficient three-dimensional aircraft recognition algorithm using normalized fourier descriptors * [J]. Computer Graphics & Image Processing, 1980, 13(2): 99-126.

[10] Chu W, Liu L. Research and Simulation Implementation of Airplane Target Typical Motion Model; proceedings of the International Conference on Computational and Information Sciences, F, 2013 [C].
[11] Wang M. Research on the Recognising the Kind and Type of A Plane by It's Characters on Movement [J]. Computer & Digital Engineering, 1999,
[12] Sain S. The Nature of Statistical Learning Theory [J]. Technometrics, 1997, 8(6): 409-.
[13] Suykens J A K, Vandewalle J. Least Squares Support Vector Machine Classifiers [J]. Neural Processing Letters, 1999, 9(3): 293-300.
[14] Huang G B, Zhu Q Y, Siew C K. Extreme learning machine: a new learning scheme of feedforward neural networks; proceedings of the IEEE International Joint Conference on Neural Networks, 2004 Proceedings, F, 2005 [C].
[15] Huang G B, Zhu Q Y, Siew C K. Extreme learning machine: Theory and applications [J]. Neurocomputing, 2006, 70(1): 489-501.
[16] Huang G B, Chen L, Siew C K. Universal approximation using incremental constructive feedforward networks with random hidden nodes [J]. IEEE Transactions on Neural Networks, 2006, 17(4): 879.
[17] Deng W Y, Ong Y S, Zheng Q H. A Fast Reduced Kernel Extreme Learning Machine [J]. Neural Networks, 2016, 76(C): 29-38.

IOP Conf. Series: Journal of Physics: Conf. Series **1061** (2018) 012016 doi:10.1088/1742-6596/1061/1/012016

Research on Obstacle Avoidance Algorithm for Four-rotor UAV

Yang Li[a], Dada Hu[b], Jingxin Chen[c] and Zhiyong Tang[d]

Beihang University, No.37 Xueyuan Road, Haidian District, Beijing, China

[a]buaa12031206@126.com; [b]510587923@qq.com; [c]buaa1203cjx@126.com; [d]57523523@qq.com

Abstract. UAV is a kind of aircraft which can be operated remotely, fly independently and carry out different tasks with special equipment on board. With the civilian application of UAV becoming more and more extensive, the application scene is becoming more and more complex. Obstacle avoidance technology has become a difficult problem in widening this field and is also the cornerstone of UAV development on a large scale. As the four-rotor UAV has the advantages of high flexibility, relatively simple control, vertical take-off and landing and low maintenance cost, in this paper, a four-rotor UAV with lidar and binocular vision camera is used to realize the obstacle avoidance function by using the artificial potential field method. The feasibility of the method is verified by simulation.

1. Introduction

UAVs are unmanned aircraft that can be operated remotely, fly independently, perform different tasks with specialized equipment on board and can be reused[1].In recent decades, with the rapid development of computer technology, artificial intelligence and control theory, the development of high-performance UAV has been widely concerned by many countries[2][3].

The development of UAV has a long history. In 1980s, the world's first UAV "Aerial Target" was successfully developed in Britain[4]. After nearly a century of development, UAVs have gradually begun to develop in different industries[5]. They can be found in the fields of remote sensing aerial photography, geological monitoring, aviation plant protection and other civilian areas as well as in military fields, such as low-altitude reconnaissance, counter-terrorism strike, border patrol and so on[6][7]. Nowadays, UAVs are developing towards chronization, stealth, combat in the air and multi-aircraft cooperation with high altitude and long voyage[8]. Because of its strong maneuverability, strong pertinence, low requirement for take-off and landing environment, low price, the small-sized UAV is being applied in all walks of life, especially as an efficient, safe and cheap way to play a great role in combating crime, military investigation and investigation of cases[9]. How to ensure that the four-rotor aircraft can safely complete the reconnaissance tasks is the key research object in the future. Therefore, it is of great significance to study the automatic obstacle avoidance technology of UAV.

In this paper, PIXHAWK flight control platform with lidar and ZED binocular camera is used as obstacle avoidance system, artificial potential field obstacle avoidance algorithm is used to avoid obstacles through the image of the obstacle distance and azimuth.

2. Structure Frame and Flight Principle of Four-rotor UAV

IOP Conf. Series: Journal of Physics: Conf. Series **1061** (2018) 012016 doi:10.1088/1742-6596/1061/1/012016

The four-rotor system consists of four-rotor aircraft and external components. The four-rotor aircraft consists of fuselage body and power system. External components include remote control, earth station, battery and other structures. The remote controls send and receive commands through the receiver of the four rotors. GPS is used to provide position signals, including longitude, latitude and altitude, and to locate.

The four-rotor aircraft is connected by two rigid straight rods of equal length with an angle of 90 degrees. When four motors are fixed at the end of the bar, the four rotor motion can be realized by properly controlling the rotation speed. In general, the reverse torsion moment between the two diagonal motors can be eliminated by the opposite direction of rotation. There are the following types of flight: vertical, pitch, roll, yaw, back and forth, lateral movement, etc.

3. Construction of flight control system

The flight control system is the core of all the systems. It collects and analyzes the attitude and position information from the sensor. After calculation, the command is given according to the instruction. In this paper, the flight control board stm32f429 (PIXHAWK) is proposed. The PX4 flight controller is used as the basic controller for motor output and attitude control. ROS robot operating system is used for task planning, obstacle avoidance and other high-level control. The communication between PX4 and ROS system is solved by using MAVLink protocol, as is shown in Figure 1.

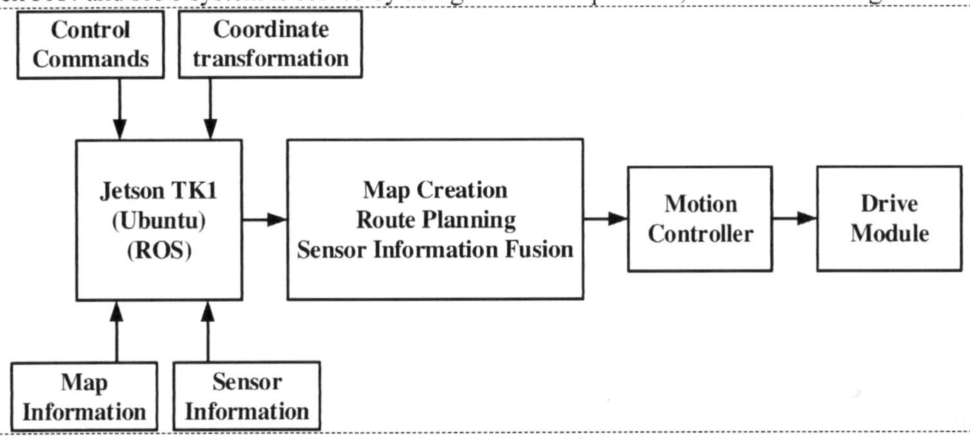

Figure 1. Integral control system.

4. Construction of obstacle avoidance module

At present, the development of UAV obstacle avoidance technology presents a multi-development state. As a whole, there are five main kinds of obstacle avoidance technology, that is ultrasonic technology, visual image composite technology, radar technology and electronic map.

In this paper, two schemes are proposed. The first is a kind of obstacle avoidance system with binocular vision. ZED stereo camera is used to obtain three-dimensional images, and accurate obstacle avoidance algorithm is used to realize the accurate perception and avoidance of outdoor obstacles. The second option can carry UTM-30LX lidar. It makes use of TOF technology and Multi-Pulse principle.

5. Obstacle avoidance algorithm

In order to realize the safe autonomous flight of four-rotor aircraft, it is far from enough to rely on obstacle avoidance technology alone. Only by combining obstacle avoidance technology with obstacle avoidance algorithm can we realize safe autonomous flight. At present, the main algorithms of UAV obstacle avoidance include gradient method, spline interpolation method, nonlinear programming method, optimal control method, A * algorithm, neural network method, simulated annealing method, genetic algorithm, ant colony algorithm, dynamic programming algorithm, etc.

In this paper, the artificial potential field method is used for path planning which is proposed by Khatib. The basic idea is to design the motion of UAV in the surrounding environment as an abstract motion in artificial gravitational field. The target point produces "gravity" to the mobile robot, and the obstacle produces "repulsive force" on the mobile robot. Finally, the motion of the mobile robot is controlled by the resultant force. The path planning using artificial potential field is generally more smooth and safe, with simpler structure and lower computational complexity and is widely applied in the field of path planning. The force analysis of UAV in artificial potential field is shown in Figure 2. The resultant force at every point in the path is equal to the sum of all repulsive and gravitational forces.

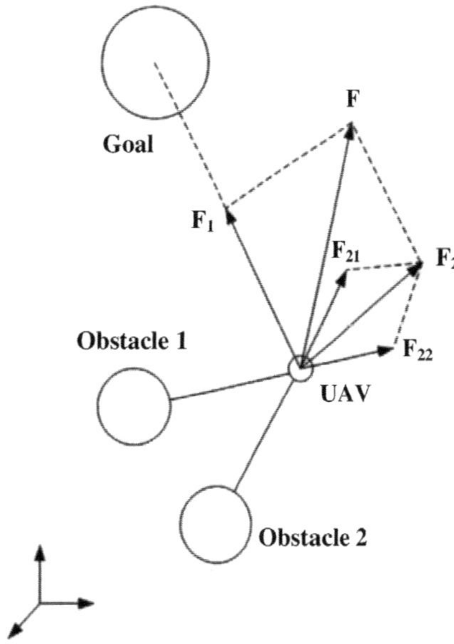

Figure 2. Force analysis of UAV.

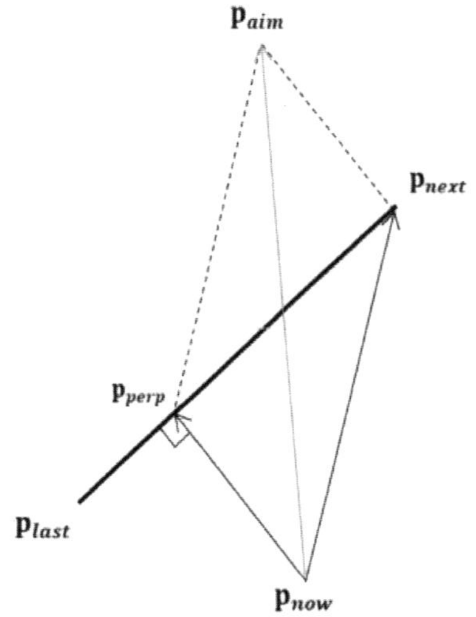

Figure 3. Stress model.

In order to realize the obstacle avoidance function of UAV, it is necessary to plan the path to make the UAV approach the planned path quickly.

If a four-rotor UAV is regarded as a particle model, then Newton's second law will be satisfied:

$$\dot{p} = \mathbf{v} \tag{1}$$

$$\dot{v} = \boldsymbol{u} \tag{2}$$

$\mathbf{p}_{now} \in R^3$ indicates the current position of the UAV, $\mathbf{p}_{next} \in R^3$ represents the target route point, $\mathbf{p}_{last} \in R^3$ represents the previous route point. A local path planner is designed to generate the desired position $\mathbf{p}_{aim} \in R^3$ in real time. The UAV is then guided to fly along the straight path between $\mathbf{p}_{last}$ and $\mathbf{p}_{next}$ to reach the target point $\mathbf{p}_{next}$, as is shown in Figure 3.

Then the foot point $\mathbf{p}_{perp}$ can be described as:

$$\mathbf{p}_{perp} = \mathbf{p}_{next} + (\mathbf{p}_{last} - \mathbf{p}_{next})\frac{(\mathbf{p}_{now}-\mathbf{p}_{next})^T(\mathbf{p}_{last}-\mathbf{p}_{next})}{||\mathbf{p}_{next}-\mathbf{p}_{last}||^2} \tag{3}$$

$$\mathbf{p}_{now} - \mathbf{p}_{perp} = \boldsymbol{A}(\mathbf{p}_{now} - \mathbf{p}_{next}) \tag{4}$$

$$\boldsymbol{A} = \boldsymbol{I}_3 - \frac{(\mathbf{p}_{last}-\mathbf{p}_{next})(\mathbf{p}_{last}-\mathbf{p}_{next})^T}{||\mathbf{p}_{next}-\mathbf{p}_{last}||^2} \tag{5}$$

Hence:

$$\mathbf{p}_{aim} = \mathbf{p}_{now} + k_1(\mathbf{p}_{next} - \mathbf{p}_{now}) + k_2(\mathbf{p}_{perp} - \mathbf{p}_{now}) \tag{6}$$

After planning path, obstacle avoidance controller can be further designed. $\mathbf{p}_{now} \in R^3$ represents the current position of UAV, $\mathbf{p}_{next} \in R^3$ represents the target point. A local path planner is designed to generate the desired position $\mathbf{p}_{aim} \in R^3$ in real time. To guide the UAV to fly to the target point $\mathbf{p}_{next}$

and avoid the obstacle with a radius of $r_0 \in R_+$ and whose center of sphere is located at $\mathbf{p}_{tree} \in R^3$. The model is shown in Figure 4.

First, two hypotheses are proposed:

1. The initial position of UAV meets the formula: $(\mathbf{p}_{now} - \mathbf{p}_{tree})^T(\mathbf{p}_{now} - \mathbf{p}_{tree}) - r_0^2 > 0$;
2. The target point $\mathbf{p}_{next}$ meets the formula: $\frac{1}{(\mathbf{p}_{next}-\mathbf{p}_{tree})^T(\mathbf{p}_{next}-\mathbf{p}_{tree})-r_0^2} \approx 0$。

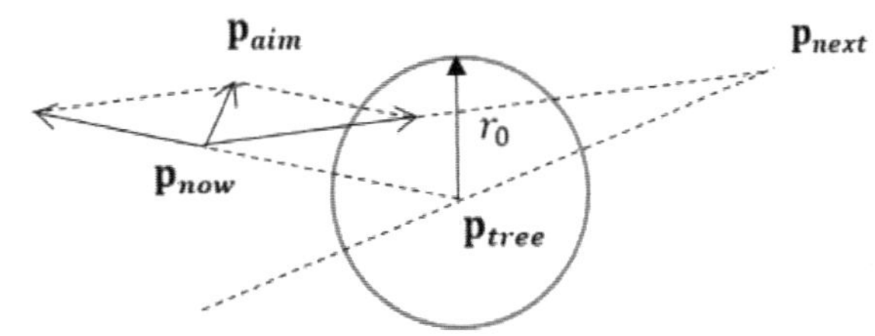

Figure 4. Obstacle avoidance model.

Then the control rate is designed as:

$$\mathbf{p}_{aim} = \mathbf{p}_{now} + k_1(\mathbf{p}_{next} - \mathbf{p}_{now}) - k_2\frac{1}{((\mathbf{p}_{tree}-\mathbf{p}_{now})^T(\mathbf{p}_{tree}-\mathbf{p}_{now})-r_0^2)}(\mathbf{p}_{tree} - \mathbf{p}_{now}) \quad (7)$$

6. Simulation

The simulation diagram of the route planning model is shown in Figure 5 :

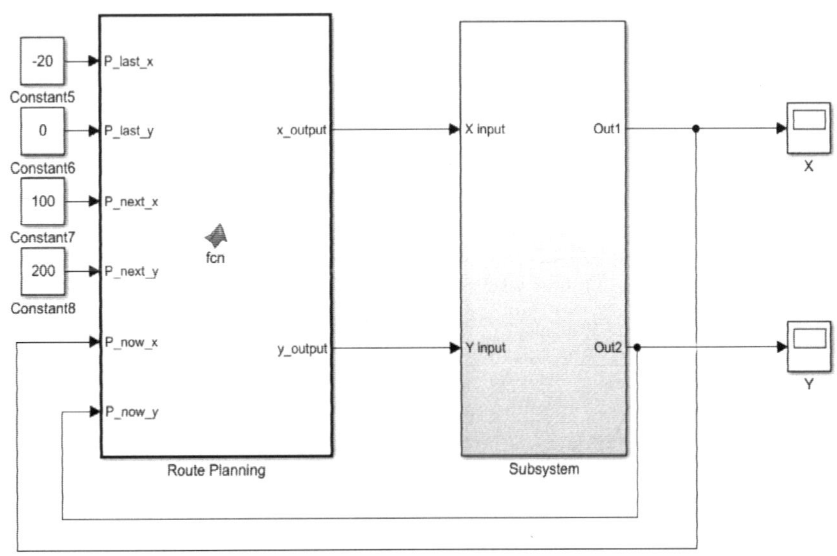

Figure 5. Route planning model.

$\mathbf{p}_{now} = (0,0)$, $\mathbf{p}_{last} = (-20,0)$, $\mathbf{p}_{next} = (100,200)$;

When $k_1 = 1, k_2 = 1$, the simulation result is shown in Figure 6:

When $k_1 = 1, k_2 = 10$, the simulation result is shown in Figure 7:

IOP Conf. Series: Journal of Physics: Conf. Series **1061** (2018) 012016 doi:10.1088/1742-6596/1061/1/012016

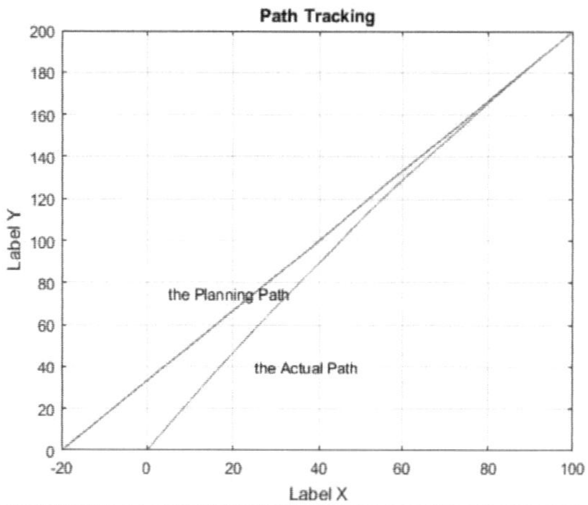

Figure 6. $k_1 = 1, k_2 = 1$.

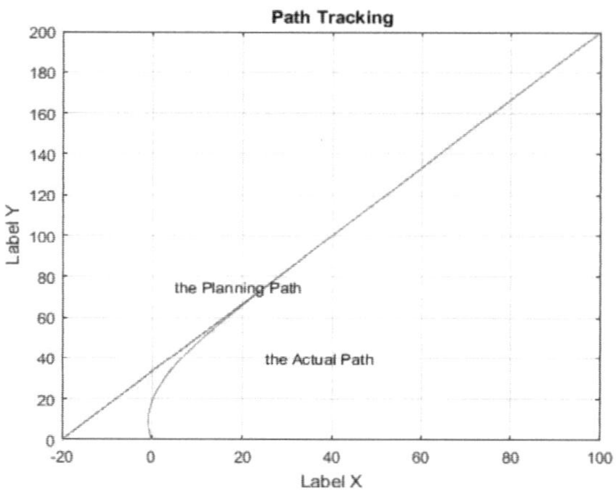

Figure 7. $k_1 = 1, k_2 = 10$

When $k_1 = 1, k_2 = 25$, the simulation result is shown in Figure 8:

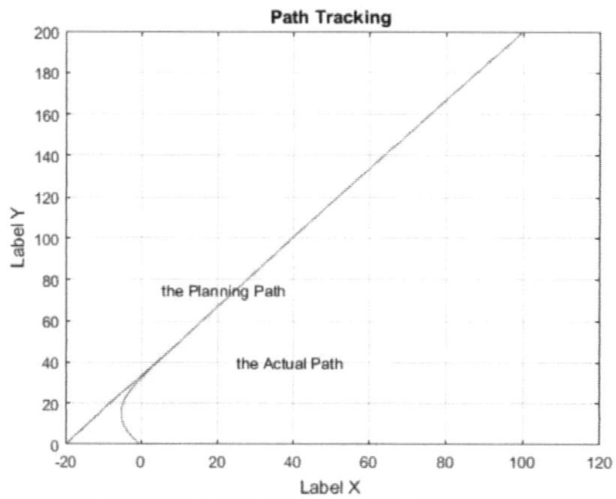

Figure 8. $k_1 = 1, k_2 = 25$

The simulation results show that: the bigger k_2 is, the faster the UAV fits the planned path.

The simulation diagram of the obstacle avoidance model is shown in Figure 9:

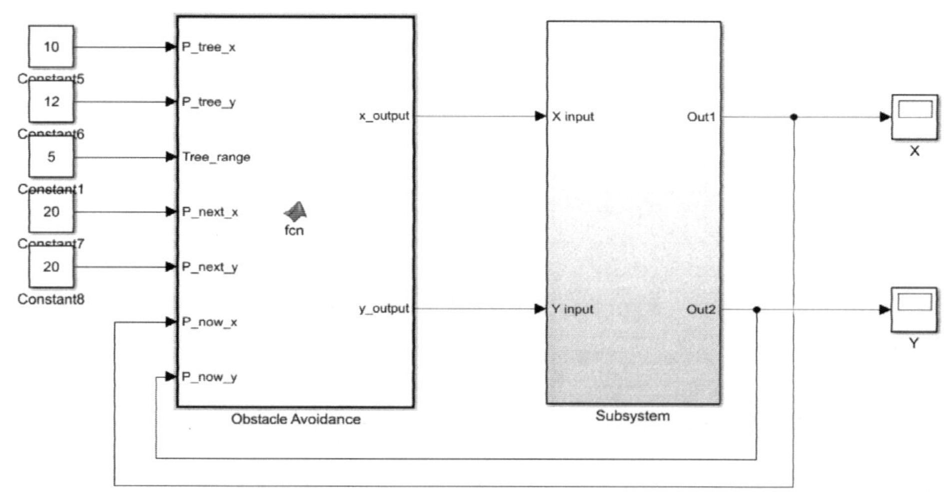

IOP Conf. Series: Journal of Physics: Conf. Series **1061** (2018) 012016 doi:10.1088/1742-6596/1061/1/012016

Figure 9. Obstacle avoidance trajectory planning model.

$\mathbf{p}_{now} = (0,0)$, $\mathbf{p}_{next} = (20,20)$, $\mathbf{p}_{tree} = (10,12)$, $r_0 = 5$;
When $k_1 = 1, k_2 = 50$, the simulation results are shown in Figure 10:
When $k_1 = 1, k_2 = 500$, the simulation results are shown in Figure 11:

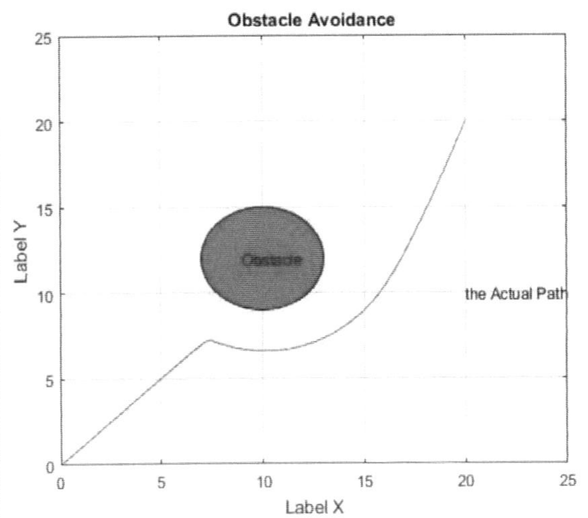

Figure 10. $\boldsymbol{k_1 = 1, k_2 = 50}$.

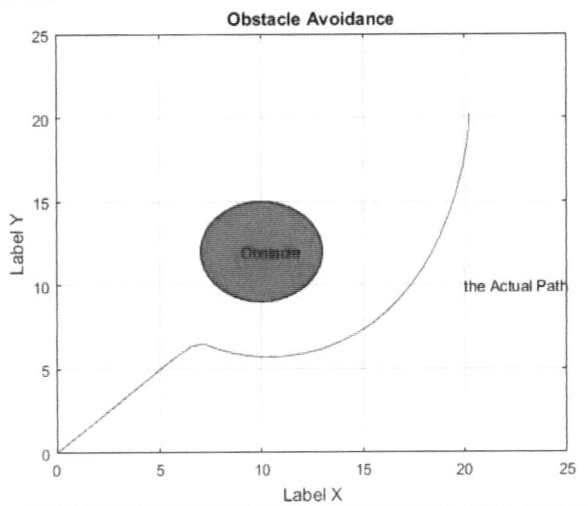

Figure 11. $\boldsymbol{k_1 = 1, k_2 = 500}$.

When $k_1 = 1, k_2 = 5000$, the simulation results are shown in Figure 12:

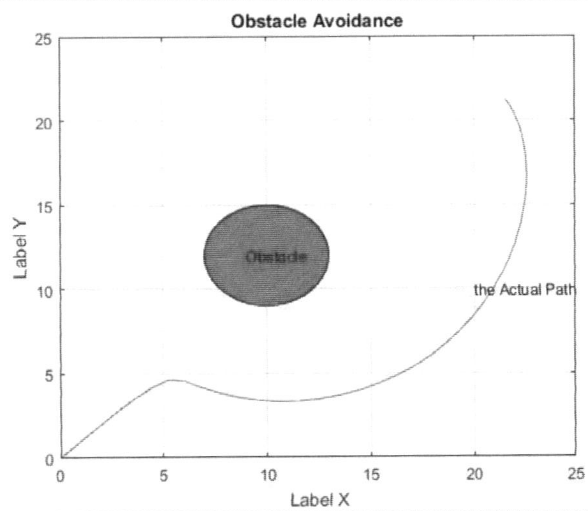

Figure 12. $\boldsymbol{k_1 = 1, k_2 = 5000}$.

According to the simulation results, when k_1 is still, the bigger k_2 is, the farther the distance from obstacle is and the better the obstacle avoidance effect is.

7. Conclusions

In this paper, a four-rotor UAV is taken as the research object. A flight control system based on lidar and binocular vision camera is built. Through the simulation of obstacle avoidance algorithm of artificial potential field, the effect is basically the same as expected, and it will be applied in the future research to the concrete experiment to verify its feasibility.

IOP Conf. Series: Journal of Physics: Conf. Series **1061** (2018) 012016 doi:10.1088/1742-6596/1061/1/012016

References

[1] Y.Chen; H.Zhang; M.Xu. The Coverage Problem in UAV Network: A Survey [C] 5th ICCCNT, 2014.
[2] L.Fu;F.Xie;D.Wang,et al. The Overview for UAV Air-Combat Decision Method [C] Control and Decision Conference (2014 CCDC), 2014.
[3] Tao Yujin, Li Peifeng. Development and key Technologies of UAV system. [J] .Aeronautical Manufacturing Technology, 2014,20: 34-39.
[4] Xu Qiyun, Wang Jie, Hao Wenyuan, et al. Developmental course and trend of foreign unmanned figurehter aircraft. [J] .Navigation missiles, 2016,3: 28-32.
[5] C. Tetrault, A Short History of Unmanned Aerial Vehicles[J/OL]. UAV Industry News, http://www.draganfly.com/news/2009/03/04/a-short-history-of-unmanned-aerial-vehicles-uavs/.
[6] Moss V, Jones D, Nwaneri S. Analysis of homeland security and economic survey using special missions unmanned aerial vehicle utilities[C]// IEEE International Geoscience and Remote Sensing Symposium. IEEE, 2012:6154-6157.
[7] Negri S, Quarta T A M. Ground penetrating radar survey for civil-engineering applications: Results from the test site[C]// International Conference on Ground Penetrating Radar. IEEE, 2012:835-839.
[8] Chun Yu Jiangmin, Zhang Heng. Development status and Prospect of UAV. [J] .Airborne missile, 2005: 23-27.
[9] Song Guanghuang. Design of a police foldable six-rotor aircraft. [J] .Science and Technology Horizon, 2014(20): 306-306.

IOP Conf. Series: Journal of Physics: Conf. Series **1061** (2018) 012017 doi:10.1088/1742-6596/1061/1/012017

Short-term Forecast for Average Speed of Road Section based on Floating Car Data with Support Vector Regression

Y H Zou, Y H Feng, J C Huang, G Q Cheng, T Wang and Z Liu

College of Systems Engineering, National University of Defence Technology, Changsha, Hunan Province, China

fengyanghe@nudt.edu.cn

Abstract. The recent intelligent transportation system has yielded remarkable progress on traffic data collection, resource allocation and intelligent programming. However, development on traffic real-time data processing and prediction still remains limited. In pursuit of real-time prediction on road section average speed, we introduced a prediction method, which mines GIS floating car data with support vector regression algorithm. The result indicated our proposed method was superior in comparison with other commonly used algorithms including linear regression, artificial neural network, Bayesian regression and ridge regression. Besides, the quick convergence and well fitting confirmed the plausibility of our method in this domain.

1. Introduction

The quick development of cities has posed great demands in building intelligent systems, which can provide us with better services in living and well manage the limited resources. Among them, the Intelligent Transportation System (ITS), targeting at optimizing the urban transportation resources, has attracted more and more attention ranging from industry, academia to government agencies [1]. Integrating information, communications, abundant sensors and other plausible technologies, ITS makes attempt to guide automatic data analysis, efficient resource allocation and smart service assignment with people, road networks and kinds of vehicles involved in [2]. Though ITS has been widely employed in city planning for a few decades, a recent significant change in ITS is revealed as the explosion of volume of data in transportations. Potential causes for this phenomenon are the rapid development of sensors for data collections and various social platforms in sharing the transportation information. A significant conclusion can be drawn that the availability of massive data is leading a revolution in the domain transferring the conventional technology-driven transportation systems to data-driven ones. To exploit disciplinary data from multi-sources and advance the performance and efficiencies in transportation systems, machine learning techniques are introduced to address related problems. A significant advantage of employment of machine learning in dealing with transportation big data lies in mining task-beneficial patterns, uncovering predictable disciplines and optimizing the process of decision-making [3]. Instead of collecting expert experience, machine learning is capable of capturing plausible rules and enhancing the generalization of systems in the era of big data [4].

With respect to ITS, there are several issues to address, such as traffic status supervision, automatically path planning, transportation vehicles allocation and etc. Among them, the prediction of real-time traffic status is regarded as one of the most critical topics, which bridges the gap between data collection and ITS design. Specifically speaking, potential information after forecasting the average speed of road sections in a real-time way can well assist the traffic control and ensure the free movement in traffic if decision-makers accommodate properly. Meanwhile, the intuitive traffic information allow more plausible strategies in choosing routines for travelers. In this paper, we would focus our attention on machine learning algorithm's employment in solving the short-term prediction of the average speed of road sections. After collecting the GIS floating car data generated from urban roads and pruning the redundant information of data, we performed calculations of the average speed of floating cars on some road section. Considering the robustness and universal generality, a classical algorithm called support vector regression (SVR) was adopted in constructing the prediction model.

The remainders of this paper are arranged as follows. Some related works are summarized in Section 2, in which we discuss methodologies of average speed prediction in road networks. Then, we detail the methods and techniques in collecting and processing the GIS floating car data as well as calculations on the average speed of cars on some road in Section 3. The SVR prediction model is introduced in Section 4 and we elaborate the principles and essence of such model. In Section 5, we collect the transportation data in Shenzhen in the form of time window to verify the plausibility and effectiveness of SVR's employment in average velocity real-time prediction.

2. Related work

The field of function estimation and regression prediction has witnessed significant progress after the adaptation of SVM method which was proposed by Vapnik [5]. A novel and efficient pairing nv-support vector regression (pair-v-SVR) algorithm was introduced by Pei-Yi Hao, which successfully combines the advantages of twin support vector regression (TSVR) and classical ε-SVR algorithms [6]. In the task of identification of nonlinear systems in RKHS spaces, SVR method has proven its effectiveness and shown the excellence in fitting residue and superiority of the regularization network in reducing computation time [7]. Considering the complex characteristics of traffic system such as nonlinearity, time-varying, randomness and uncertainty, Thomas Epelbaum et al. utilized deep learning models to capture regression disciplines in time series data [8]. These algorithms are designed to address real-time average speed prediction of road section based on Floating Car Data (FCD).

The former works with respect to traffic flow characteristics focus mainly on relationships between three traffic flow characteristics as traffic flow, average speed and density. Greenshields firstly developeda linear model describing the relationship between velocity and density in 1993 [9]. The model, which was adopted by the U.S. Department of Transportation, assumed that the flow rate is linear with the velocity before it reaches to the maximum and then is illustrated in curve relation when the flow rate is between the maximum and the coordinate origin point [10]. Natalia Isaenko et al. designed an integrative framework which was capable of recognizing and selecting suitable method for traffic forecasting with individual FCD [11].

The main challenge of analyzing the FCD comes from the geographic data error. Map-matching is an important step in the information processing that can minimize errors effectively. Jia-Ching Ying et al. developed a novel modularity-based map-matching algorithm called Urban Map-Matching (UrbMatch) utilizing urban GPS trajectories [12]. The method called spatial and temporal conditional random field (ST-CRF) has better performance and robustness when facing the low-frequency trajectory data(e.g., one GPS point for every 1–2 minutes) [13]. Mahdi Hashemi put forward a weight based map-matching algorithm, which can be applied in real-time complex urban road networks [14]. In order to calculate the average speed of road sections, Yanace J. L. et al. performed least square method on the instantaneous speed of the floating car [15]. However, from the perspective of statistics, it is difficult to control the estimation error.

3. Floating vehicle data analysis and processing

3.1. Floating vehicle sample collection

Floating car, also called probe vehicle, refers to a vehicle equipped with a GPS positioning system and a wireless communication device. They can collect their own traffic data, such as speed, transmission time, latitude and longitude, direction, passenger status, the distance between the last point and other information on roads. The collection of FCD is a sampling survey process on the road traffic network.

The number of floating car samples should be determined before the collection of floating car traffic information. The quantity of floating cars in the road network should be big enough to ensure the accuracy of traffic flow parameter estimation. At the same time, however, the relationships between road coverage, the information update cycle and the number of floating vehicles should be furthr discussed. The former works [16, 17] provide ways to determine the number of floating car samples, with multiple factors considered. The relationship between the number of floating car samples, traffic parameters, road coverage, and information refresh cycle can be approximated as follows:

$$\beta = \left(1 - \mathrm{e}^{-Nk}\right)^2 + Nke^{-2Nk} \quad (1)$$

Where N is the number of floating car samples; β is the road coverage; K is the floating car density:

$$K = \frac{\overline{v}t_0}{l} \quad (2)$$

Where, $\overline{v}$ is the average speed of traffic flow; t is the information update cycle and l is the length of the link.

Given the length of the link, the information refresh cycle, and the traffic flow rate, the relationship between coverage and sample size can be obtained.

3.2. Floating car data processing

After accessing the data of the floating car, it is generally the first step to match the map and prune some anomaly data. Map-matching is quite crucial for floating car information processing. The intrinsic ideology is to compare the vehicle locating trajectory obtained by the data acquisition system with the road information in the electronic map database, and then map the vehicle to the most probable position on the map by some effective algorithm [18]. In this paper, we made use of the ST-Matching algorithm to embed the information of spatial connection in the road network. The Figure 1 illustrates the variation after matching crossroads data point with ST-matching algorithm.

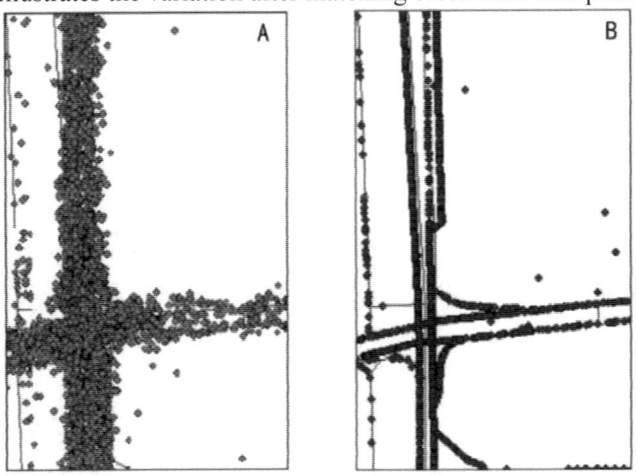

Figure 1. Map-matching of floating car data points. (A) The raw location data on the map. (B) Map matched data points are on the most probable position of the road line.

In order to eliminate the invalid data, the criteria for determining invalid data should be determined. Basing on the requirements of predictive modeling and the actual situation, this paper identifies the following screening criteria:

- The wrong data.

Repeated records. The data in one car ID which have different positions but the same time. The data of two neighbor points with the same car ID which have different positions but the distance is 0.

- The invalid data

Considering the taxis without passengers tend to find the business at a low speed or parking, their data cannot reflect the real situation. If a same ID car in the same latitude and longitude for a long time (2 *min* above), we regard these vehicles in abnormal driving conditions, which data should be removed. If two neighbor floating car points have a too-long (120 seconds above) time interval, it cannot truly reflect the real average speed. Due to the speed limit on urban roads, the calculated car speed should satisfy the constraint. The data whose average speed is more than 80km/h should be removed [19].

4. Support vector regression model

Two typical SVR algorithms, ε-SVR and ν-SVR, are commonly used in regression. ν-SVR is superior to ε-SVR in terms of regression accuracy and maneuverability [20]. In order to avoid the artificial error in the choice of coefficients, we chooses ν-SVR as the prediction model. One of advantages in SVR is the powerful feature transformation trick which maps the input into high dimensional Hillbert space. The nonlinearity of the mapping empowers better representational capability of characterizing the potential decision boundary. Besides, the model is embedded with the tolerant mechanism which makes the model more stable even in the presence of response noise. We provide brief description of ν-SVR formulation following Vapnik as follows [21]:

Given the training dataset $T = \{(x_i, y_i)|i = 1,2,\ldots, N\}$ where the x is the attribute vector while the y is the response variable, the objective of SVR is to capture the decision function in the form as $f(x) = w^T * \phi(x) + b$. It can be noticed the decision function is accompanied with the potential feature transformation which can be learned in a parameter tuning process.

$$\min_{w,b,\epsilon^{(*)},\epsilon} \frac{1}{2}\|w\|^2 + C * (\nu\varepsilon + \frac{1}{N} \Sigma_{i=1}^{N}(\xi_i^* + \xi_i)) \tag{3}$$

s.t.

$$|(w^T * \phi(x) + b) - y_i| \leq \xi_i^* + \xi_i, i = 1,2,..,N$$
$$\xi_i^{(*)} \geq 0, i = 1,2,..,N$$
$$\xi_i \geq 0, i = 1,2,..,N$$

The hyper parameter C controls the extent of tolerance to the noise and the norm of parameter, w can be viewed as the penalty term which generally leads better generalization [22]. The count of support vector is controlled by the parameter ν, in range (0, 1]. Note that training samples falling inside the ε-tube have zero loss, and samples outside the ε-insensitive zone are linearly penalized using the slack variables $\xi_i^*, \xi_i \geq 0$, i = 1,…,N. Usually, problem(3) is usually solved in its Lagrange's dual form, (see [23] for details):

$$\max_{a^*} \frac{1}{2} \sum_{i,j=1}^{l} (a_i^* - a_i)(a_j^* - a_j)K(x_i, x_j) + \sum_{i=1}^{l} y_i(a_i^* - a_i) \tag{4}$$

s.t.

$$\sum_{i=1}^{l} (a_i^* - a_i) = 0$$
$$0 \leq a_i^* \leq \frac{C}{N}, \quad i = 1, \ldots, N$$
$$\sum_{i=1}^{l} (a_i^* - a_i) \leq C\nu$$

Where the kernel $K(x_i, x_j) = (\Phi(x) \cdot \Phi(x'))$.

Solution of the dual formulation (4) yields optimal values of parameters $\overline{a}^{(*)} = (\overline{a}_1, \overline{a}_1^*, \ldots, \overline{a}_N, \overline{a}_N^*)$ that can be used to construct the optimal SVR function: $f(x) = \sum_{i=1}^{N}(a_i^* - a_i)(x_i, x_j) + \overline{b}$ In this

IOP Conf. Series: Journal of Physics: Conf. Series **1061** (2018) 012017 doi:10.1088/1742-6596/1061/1/012017

optimal solution, training samples with non-zero coefficients are the support vectors (SVs), corresponding to data points at the boundary or outside ε-insensitive zone.

The non-linear kernel $K(x_i, x_j)$can be computed with the dot product (x_i, x_j) in (7) to extend linear SVR to a non-linear setting. This kernel $K(x_i, x_j) = (\Phi(x) \cdot \Phi(x'))$ implicitly captures the non-linear mapping of the data $x \rightarrow \varphi(x)$.

Some commonly used kernel functions includePolynomial kernel, Radial Basis Function (RBF), and Sigmoid kernel [24].

The prediction index of the model is the average speed of the target road section, as well as its upstream and downstream sections, over the past period of time. To construct the input matrix, we assume the average speed of road l on time t is $\overline{V}_l(t)$ and the input data format is:

$$\begin{bmatrix} \overline{V}_l(t+1-p) & \overline{V}_l(t+2-p) & \cdots & \overline{V}_l(t) \\ \overline{V}_{l+1}(t+1-p) & \overline{V}_l(t+2-p) & \cdots & \overline{V}_{l+1}(t) \\ \vdots & \vdots & & \vdots \\ \overline{V}_{l+k}(t+1-p) & \overline{V}_{l+k}(t+2-p) & \cdots & \overline{V}_{l+k}(t) \end{bmatrix}$$

Where p is the number of periods (1 minute) that needs to be traced back when predicting the average speed. The data is divided into training set and testing set to carry out related experiments.

5. Experiment

In this section, we will employ the taxi float car data of Futian District, Shenzhen on October 1st, 2017 from 17:00 to 19:00 in the experiment. After the data preprocessing, we compared SVR predicted results with results using other commonly used regression prediction algorithms.

5.1. Data Processing

We select the average speed data of the unidirectional section of Hongli Road from east to west (from Caitian Road to Xintian Road). The parameters of the target section and its related sections of the attributes shown in Table 1.

Table 1. Parameters of road sections.

	Name	Num.	Range of Rd.	Length(km)
Upstream	Hongli Rd.	5	Jintian Rd.-Caitian Rd.	0.490
sections	Jintian Rd.	4	Fuzhong Rd.- Hongli Rd.	0.428
Target section	Hongli Rd.	1	Yitian Rd.- Jintian Rd.	0.702
Downstream	Hongli Rd.	2	Yitian Rd.-Xinzhou Rd.	0.705
sections	Yitian Rd.	3	Fuzhong Rd.- Hongli Rd.	0.434

The target road section has two upstream sections (No.4, No.5) and two downstream sections (No.2, No.3), which have direct influence to the target downstream. Cars on the target section all come from upstream sections, and drive to downstream sections.

We first run a map-matching process and then the scattered points are projected onto the corresponding line segments of the road based on the driving direction of the vehicle. The speed and driving direction of the vehicle can be calculated by the change of the latitude and longitude.

According to the calculation formula mentioned in Section 3.1, the road coverage rate is calculated as follows:

Table 2. Road Coverage Rate Calculation.

Rd. Number	1	2	3	4	5
Length(km)	0.702	5	0.434	0.428	0.490
Lowest speed(km/h)	14.27973	14.27973	4.083369	9.015247	6.585276

k	1.695125	1.687911	0.784057	1.755305	1.119945
The minimum of samples N	6	10	4	7	5
road coverage β	0.999923	1	0.920915	0.999991	0.992693

We define the length of a time interval as 5 minutes and the road coverage rate of each road segment are calculated under the minimum sample quantity and the lowest average speed. The minimum number of samples is collected from the time interval when there are least FCD points on the road. As can be seen in the Table 2, the road coverage rate can reach more than 90% with the minimum of average speed and the minimum of samples. In this case, the sample size in any time interval can satisfy the minimal accuracy requirement.

5.2. Prediction model

According to the requirements of SVR model training and the characteristics of road sections, the average speed of the road sections needs to be prepared in the following format.

Table 3. Data Format of the Input and Output.

							Prediction Index					
Rd. Number	1			1		2	3	4			5	
Sample	V_{t+1}	V_t	V_{t-1}	$\cdots$	V_{t-p+1}	$\cdots$	$\cdots$	$\cdots$	V_t	V_{t-1}	$\cdots$	V_{t-p+1}
	Output						Input					

Where p is the backtracking coefficient.

The calculated data of the various road sections are represented in a matrix according to the format in Table 4, and the data is processed by setting p = 3 as an example first, a matrix of [111 × 16] is formed. Then, we perform normalization on these data. Finally, we partitioned the first 80 lines as training set and the last 31 lines as testing set.

In the process of model training, the parameter C of the RBF kernel ranges in the interval [-5,200], with a step of 5, and the parameter g of the kernel ranges in the interval [-5,5], step size is 1. We performed grid search on these parameters to obtain the optimal parameters. The result is on Figure 2:

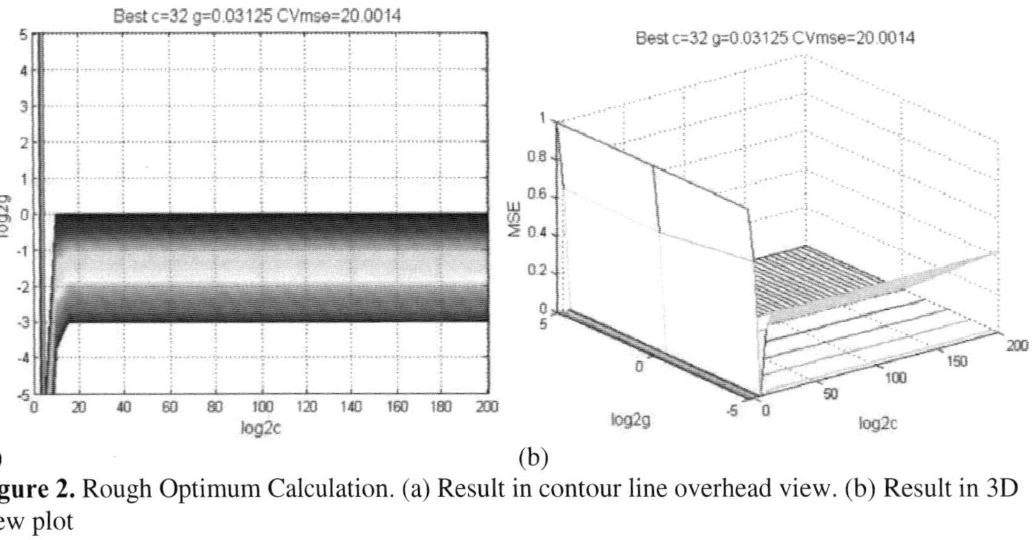

(a) (b)

Figure 2. Rough Optimum Calculation. (a) Result in contour line overhead view. (b) Result in 3D view plot

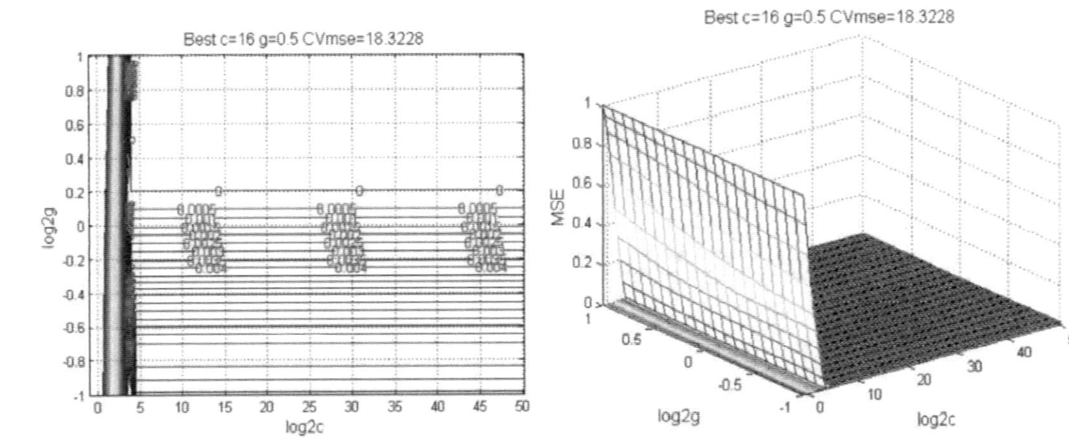

(a) (b)

Figure 3. Accurate Optimum Calculation. (a) Result in contour line overhead view. (b) Result in 3D view plot.

Basing on the result of the rough selection of the parameters, we narrow the search range. We set the parameter C, in the interval [-1, 50], with a step of 0.5, and the parameter g, in the interval [-1,1], with a step length of 0.1, take an accurate optimum calculation. The result is on Figure 3.

According to the optimization results, the best *C* parameter is 16, *g* is 0.5. The mean square error at this time is 18.3228. The forecast results are shown in Figure 4

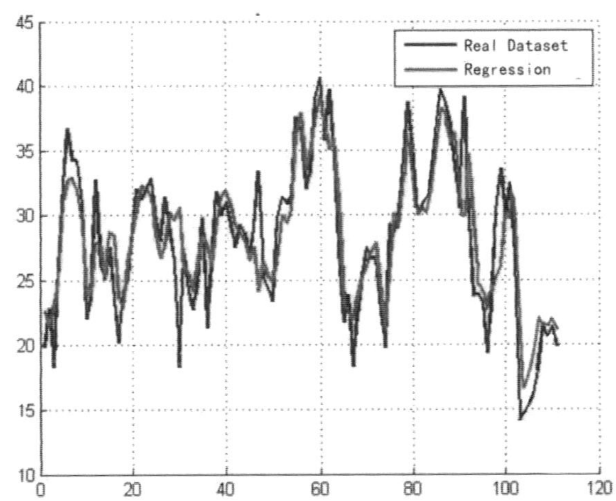

Figure 4. SVM regression curve on the data set. Training set (before 80) and the testing set (after 81).

To better evaluate the performance of the model, we compared the predicting results of SVR, ANN, linear regression, Bayesian ridge and ridge regression on the same dataset. The results shows on Figure 5, The MSE of each algorithm results on the testing set shows on Table 4.

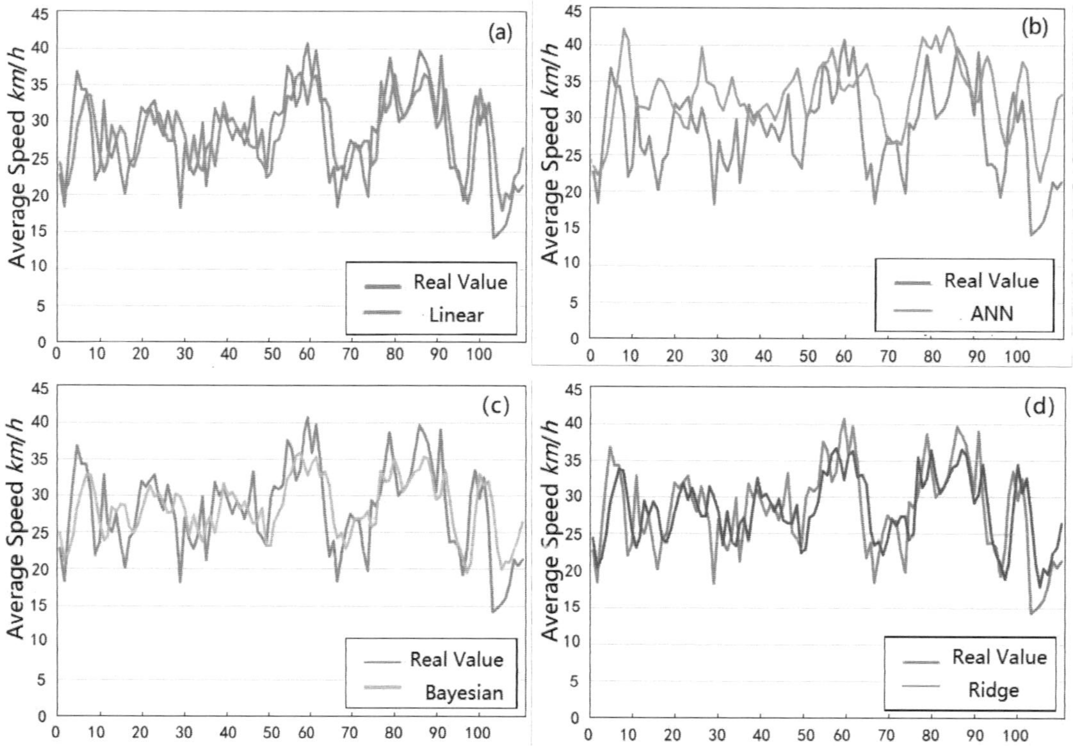

Figure 5. Comparison of the fitting degree on the dataset. (a) Linear regression. (b)ANN. (c) Bayesian ridge. (d) Ridge regression.

Table 4. The MSE of each algorithm results on the testing set.

Algorithm	SVR	Linear model	ANN	Bayesian ridge	Ridge
MSE	18.3228	22.1685	71.2724	26.0315	22.1801

6. Discussion and conclusion

In this work we demonstrated the superior performance of SVR on the average speed of the road section regression forecast. Instead of considering the data from a single road section, we aggregate the data of related road sections in the high dimensional space, aiming at achieving quick convergence. Our results suggest that SVR can handle this type of input well, with a smaller mean square error than the other algorithms.

The condition considered in this work is a typical scene on the urban road network. SVR can deal with the forecast problems under longer backtracking time and the complex road network conditions. We can find more regular patterns in the daily long-term data of each road section in our forecast index system, when accessing more data. In future work, we would address these limitations, adapting our forecast method to accommodate the daily long-term and road network domains.

Acknowledgment

Thanks to Dr. Qi Wang's help in academic wrting. Yehan Zou acknowledges Support by National Natural Science Foundation of China (Grant No: 71701205,71701206).

References

[1] Shivendra, A. and S. Kumarswamy, Intelligent Transportation System. Computer-Aided Civil and Infrastructure Engineering, 2016. 18(3): p. 173-183.

[2] Souza, A.M.D., et al. Real-time path planning to prevent traffic jam through an intelligent transportation system. in Computers and Communication. 2016.
[3] Jha, V., Study of Machine Learning Methods in Intelligent Transportation Systems. 2015.
[4] Wang, Q., et al., Addressing Complexities of Machine Learning in Big Data: Principles, Trends and Challenges from Systematical Perspectives. 2017.
[5] Vapnik, V., S.E. Golowich, and A. Smola, Support Vector Method for Function Approximation, Regression Estimation, and Signal Processing. Advances in Neural Information Processing Systems, 1996. 9: p. 281--287.
[6] Hao, P.Y., M.S. Lin, and L.B. Tsai. A New Support Vector Machine with Fuzzy Hyper-Plane and Its Application to Evaluate Credit Risk. in Eighth International Conference on Intelligent Systems Design and Applications. 2008.
[7] Sayehi, I., et al. A comparative study of two kernel methods: Support Vector Regression (SVR) and Regularization Network (RN) and application to a thermal process PT326. in International Conference on Sciences and Techniques of Automatic Control and Computer Engineering. 2016.
[8] Epelbaum, T., et al., Deep Learning applied to Road Traffic Speed forecasting. 2017.
[9] Arnott, R., E. Inci, and J. Rowse, Downtown curbside parking capacity. Journal of Urban Economics, 2015. 86: p. 83-97.
[10] Transportation, U.S.D.o., Chapter 3. Numerical Modeling. 2008.
[11] Isaenko, N., C. Colombaroni, and G. Fusco. Traffic dynamics estimation by using raw floating car data. in IEEE International Conference on MODELS and Technologies for Intelligent Transportation Systems. 2017.
[12] Ying, J.C., et al., Spatial-temporal Mining for Urban Map-Matching.
[13] Liu, X., et al., A ST-CRF Map-Matching Method for Low-Frequency Floating Car Data. IEEE Transactions on Intelligent Transportation Systems, 2017. 18(5): p. 1241-1254.
[14] Hashemi, M. and H.A. Karimi, A weight-based map-matching algorithm for vehicle navigation in complex urban networks. Journal of Intelligent Transportation Systems, 2016: p. 00-00.
[15] 15.Ygnace, J.L., et al., Travel Time Estimation on the San Francisco Bay Area Network Using Cellular Phones as Probes. 2000.
[16] Zhi, T.U., et al., Study on the Route Coverage and the Update Cycle of Transportation Information Based on the Minimum Samples of Floating Car. China Railway Science, 2006. 27(5): p. 127-131.
[17] Chen, Y., Study on the density of floating car based on the route coverage. Journal of Shandong University of Technology, 2006.
[18] Jensen, C.S. and N. Tradišauskas, Map Matching. Encyclopedia of Database Systems, 2016: p. 1692-1696.
[19] Jahnke, M., et al., Identifying Origin/Destination Hotspots in Floating Car Data for Visual Analysis of Traveling Behavior. 2017: Springer International Publishing.
[20] Miguel, M. and A.T. Flor, Spectroscopic Determination of Aboveground Biomass in Grasslands Using Spectral Transformations, Support Vector Machine and Partial Least Squares Regression. Sensors, 2013. 13(8): p. 10027.
[21] Vapnik, V., Estimation of Dependences Based on Empirical Data. 2006: Springer New York.
[22] Wang, Q., et al., A Novel Ensemble Method for Imbalanced Data Learning: Bagging of Extrapolation-SMOTE SVM. Computational Intelligence & Neuroscience, 2017. 2017(3): p. 1827016.
[23] Cherkassky, V., Predictive Learning, Knowledge Discovery and Philosophy of Science. 2012: Springer Berlin Heidelberg. 209-233.
[24] Smola, A. and V. Vapnik, Support vector regression machines. 1997.

IOP Conf. Series: Journal of Physics: Conf. Series **1061** (2018) 012018 doi:10.1088/1742-6596/1061/1/012018

Simulation and Design of Solar Power System for Ocean Buoy

Jingxin Chen[a], Yang Li[b], Xuen Zhang[c] and Yaofei Ma[d]

Beihang University, No.37 Xueyuan Road, Haidian District, Beijing, China.

[a]buaa1203cjx@126.com; [b]buaa12031206@126.com; [c]13240315291@163.com; [d]mayaofeibuaa@163.com

Abstract. Promoting maritime power and improving the ability of planning ocean is considered to be an important issue of nation security and development in China, also a pivotal to implement 'The Belt and Road' strategy. As an automatic observation station on the sea, ocean buoys play a vital part in this strategy. The majority of ocean buoys, powered by accumulator, work far away from the shore. Thus, solar photovoltaic system is a valid supply power for ocean buoys, which can greatly improve the battery life and make ocean buoys more economical. This essay contraposes the problem of continuous power supply for ocean buoys working on the sea and design a solar photovoltaic system for ocean buoys that works on the South China Sea providing power of 1kW and voltage of 36V and also a protection of 72 hours supply on extreme occasions. In the end, there will be some expansion about other cities on their solar energy resources and solar photovoltaic system parameters.

1. Introduction

It is our responsibility and opportunity to vigorously carry out "The Intelligent Ocean" project to help implement the national maritime strategy and further enhance the informatization level of the marine industry. Offshore ocean buoys are a very important part of the ocean exploration. It is an automatic observation station that floats for a long time on the sea.

In other countries, research and development on ocean buoy began in the late 1940s and early 1950s. Ocean buoy was firstly applied to marine surveys in the 1960s. In the 1970s, ocean buoy entered a practical phase and the technology became more and more mature. Later on, there were more and more types of ocean buoys and fields of application became more and more comprehensive. Because of the continuous development of technologies such as micro-processing technology and satellite communication technology, the working capability of ocean buoy is getting stronger and more stable. At present, ocean buoy technology develops quickly in the United States, Japan, France, the Soviet Union, the Netherlands, Australia, Canada etc. In the middle of 1960s, China began to research and develop ocean buoys. Since the 1990s, the actual application of ocean buoys in our country has officially started. After promoted by the "Seventh Five-Year" national project, ocean buoy technology has been maturing and improving, which is gradually catching up with the United States, Japan and other countries that have an earlier and faster development in ocean buoys.

In general, ocean buoys are located offshore from land and cannot be interconnected with the public power grid on land. To power the ocean buoys and reduce the number of batteries used, a suitable power supply system is required. Solar power system (photovoltaic power generation system) is a good choice due to abundant solar energy and easy collection. Using solar energy to charging batteries, the number of replacement batteries can be greatly reduced, thereby cost reduced.

IOP Conf. Series: Journal of Physics: Conf. Series **1061** (2018) 012018 doi:10.1088/1742-6596/1061/1/012018

Optimizing the angle of the solar panel slope can further allow the system to accept as much solar radiation as possible. In July 2015, Baidu Cloud Computing (Yangquan) Centre's solar photovoltaic power generation project was successfully connected to the grid. This is the first application of solar photovoltaic technology in domestic data centers [1].

2. Design of ocean buoy's solar power system configuration scheme

2.1. Design principle

In the design and configuration of solar power system, it is necessary to meet the demand of the load under the conditions of safety. Improving reliability and reducing cost are also important. For a stand-alone solar power system, the design principle is to meet the load's demand under different solar irradiances in different seasons. Because the solar radiation in winter is relatively weak [2], system designed should at least meet the load's demand in winter.

2.2. Structure of solar power system and functions of each part

In this project, solar power system consists of the following four main parts:

- Photovoltaic cell array (solar panel)
- Storage Battery pack
- Photovoltaic controller
- Inverter

Structure of this system is shown in figure 1.

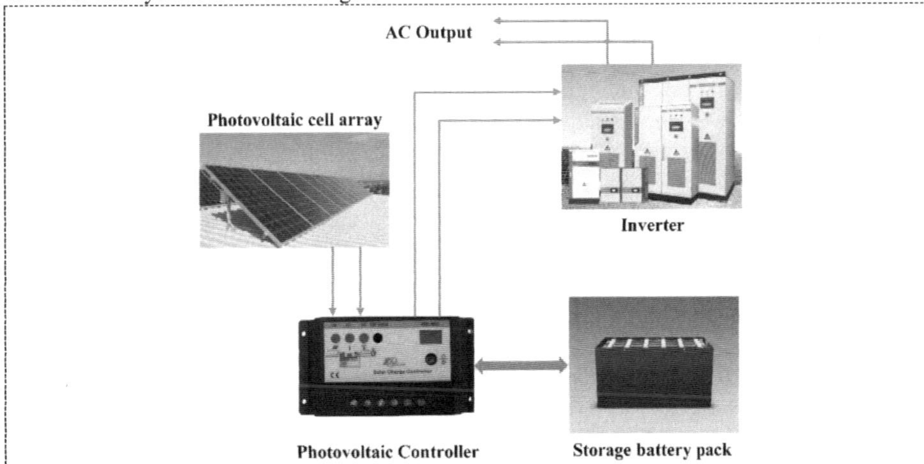

Figure 1. Structure of solar power system.

The functions of each part are as follows:

- **Photovoltaic cell array (solar panel)**[3]. Photovoltaic cell array consists of photovoltaic batteries in series and parallel, which is often seen in everyday life of solar panels. This is a very important core part of solar power system, ensuring that solar energy is converted into electrical energy. Electrical energy can be stored in storage battery pack. It can also be directly used to charge the buoy.
- **Photovoltaic Controller**. Photovoltaic controller is located in the unidirectional current channel between the photovoltaic cell array and the inverter, and it is bi-directionally connected with the storage battery pack to control the working status of the system, that is,
  a) The ocean buoy uses electrical energy from photovoltaic cell array.
  b) The ocean buoy uses electrical energy from storage battery pack.
  c) Provide over-charge protection and over-discharge protection for storage battery pack.

When the temperature difference between day and night in a large area, the controller can also provide temperature compensation. In addition, the PV controller also has many other features, such as the choice of light control switch or time switch, the system provides with electronic short circuit protection, overload protection or unique anti-reverse protection.

- **Storage battery pack**. As the name implies, this part can store electrical energy, in case that there is less or no solar energy in winter, at rainy days or at night, when photovoltaic cell array cannot produce electricity.
- **Inverter**. It can change DC power into AC power.

3. Modelling and calculation mechanism

3.1. Solar radiation's calculation model

At present, the calculation models of incline irradiation mainly include the following:

- Proposed by Liu and Jordan, improved by Klein[4][5], and based on that solar has the same characteristic in isometric isotropy, this algorithm considers the scattered radiation and the ground-reflection radiation in isometric isotropy. The method is relatively simple but not accurate enough, which only has relatively high accuracy in spring equinox and autumnal equinox.
- Klin Klein and Theilacker proposed the average lunar bf illumination method (KT) in 1981, the algorithm shows anisotropy of solar energy in different direction in the sky, which improves the accuracy but increases complexity. [6]

The most important mathematical model above is KT proposed by Klein and Theilacker. This method can calculate the irradiance of the array slant plane with different latitudes, different orientations and different inclinations. In addition, Klein further analysed the calculation results and proposed a new quaternary method to adjust the inclination change period, which can further increase the annual irradiance obtained on the array surface.

In order to improve the accuracy of solar radiation calculation, KT method will be applied.

4. Simulation

4.1. User Interface Introduction

Simulation system in this paper achieves the aims below:

- Solar radiation modelling based on sky anisotropy model (KT method)
- Configuration of system capacity
- Optimization based on bevel inclination

Use MATLAB Graphical User Interface (GUI), these parts will be presented with a selection of interfaces, the interface is saved as a *.m* file named chooseplease. Open it and click the corresponding button, you can achieve the corresponding model program running.

4.2. Solar radiation calculation simulation

Solar radiation modelling based on sky anisotropy model (KT method) is shown in figure 2 and figure 3.

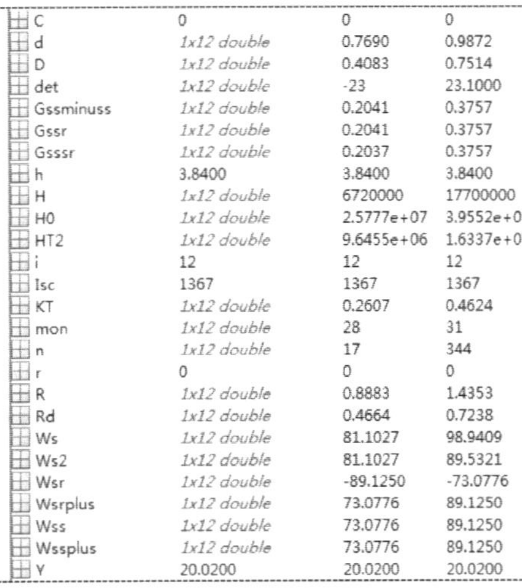

C	0	0	0
d	1x12 double	0.7690	0.9872
D	1x12 double	0.4083	0.7514
det	1x12 double	-23	23.1000
Gssminuss	1x12 double	0.2041	0.3757
Gssr	1x12 double	0.2041	0.3757
Gsssr	1x12 double	0.2037	0.3757
h	3.8400	3.8400	3.8400
H	1x12 double	6720000	17700000
H0	1x12 double	2.5777e+07	3.9552e+07
HT2	1x12 double	9.6455e+06	1.6337e+07
i	12	12	12
Isc	1367	1367	1367
KT	1x12 double	0.2607	0.4624
mon	1x12 double	28	31
n	1x12 double	17	344
r	0	0	0
R	1x12 double	0.8883	1.4353
Rd	1x12 double	0.4664	0.7238
Ws	1x12 double	81.1027	98.9409
Ws2	1x12 double	81.1027	89.5321
Wsr	1x12 double	-89.1250	-73.0776
Wsrplus	1x12 double	73.0776	89.1250
Wss	1x12 double	73.0776	89.1250
Wssplus	1x12 double	73.0776	89.1250
Y	20.0200	20.0200	20.0200

Figure 2. Parameters in the work area.

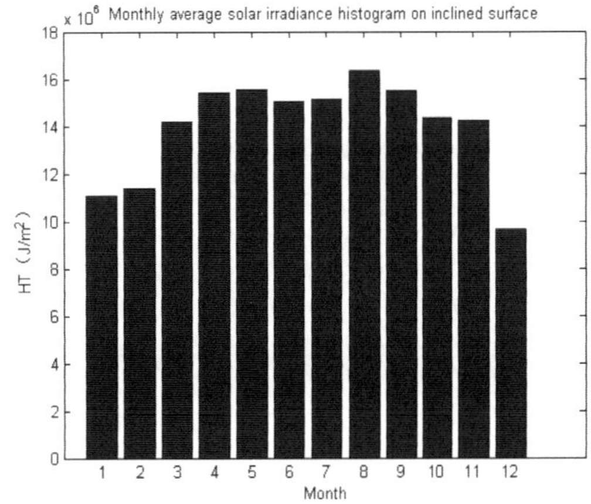

Figure 3. Average monthly solar total radiation

4.3. Configuration of photovoltaic cell array and storage battery pack

Take Haikou City as an example.

In the simulation system, solar energy system's capacity configuration is realized through programming, the results are as follows.

250W polysilicon photovoltaic module is adopted. Its parameters and configuration are shown in table 1.

Table 1. Configuration of photovoltaic cell array.

Rated power (W)	Open circuit voltage (V)	Peak operating voltage (V)	Short circuit current (A)	Peak operating current (A)	Serial number	Parallel number	Total power (W)	Total Weight (kg)
250	37.4	30.3	8.7	8.08	2	31	15179.088	1080

Parameters and configuration of storage battery adopted is shown in table 2.

Table 2. Configuration of storage battery pack.

Specification	Standard voltage (V)	Nominal capacity (Ah)		Reference Dimensions (mm)				Reference weight (kg)	Serial number	Parallel number	Total Weight (kg)
		10h	1h	Length	width	height	total height				
JGFM-1200	2	1200	780	475	174	328	365	79	18	3	4266

4.3.1. Photovoltaic controller and inverter. Parameters of photovoltaic controller adopted is shown in table 3

Table 3. Photovoltaic controller.

Specification	Rated voltage	Rated current	Number of input of PV array
SD48300	DC48	300	6

Parameters of inverter adopted is shown in table 4

Table 4. Inverter.

Specification	Input rated voltage (V)	Input rated current (A)	Rated output power (kW)	Range of input voltage (V)	Inverter efficiency
SN482KS	DC48	48	1.6	42~64	86%

4.4. Simulation optimization on inclination of photovoltaic cell array s' bevel

Take Haikou City as an example.

The MATLAB program is further improved, so that the inclination ***bb*** in the program is changed from 0 to 90 in a certain step, and the results of each angle value are compared. The main method used to generate a graph to compare irradiation parameters is shown in figure 4 and figure 5.

How solar radiation in winter varies with inclination is shown in figure 4.

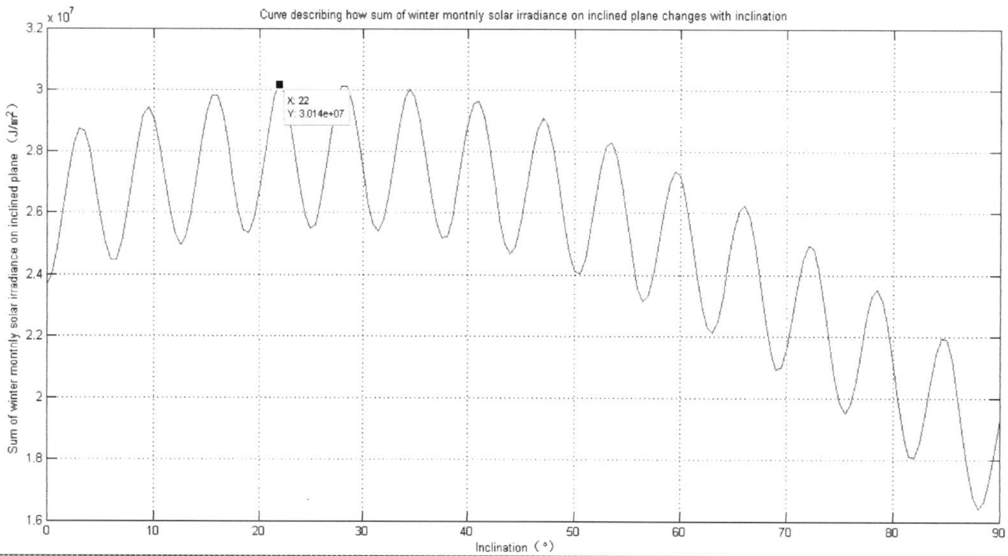

Figure 4. Slope of solar radiation in winter varying with inclination.

How solar radiation of the whole year varies with inclination is shown in figure 5.

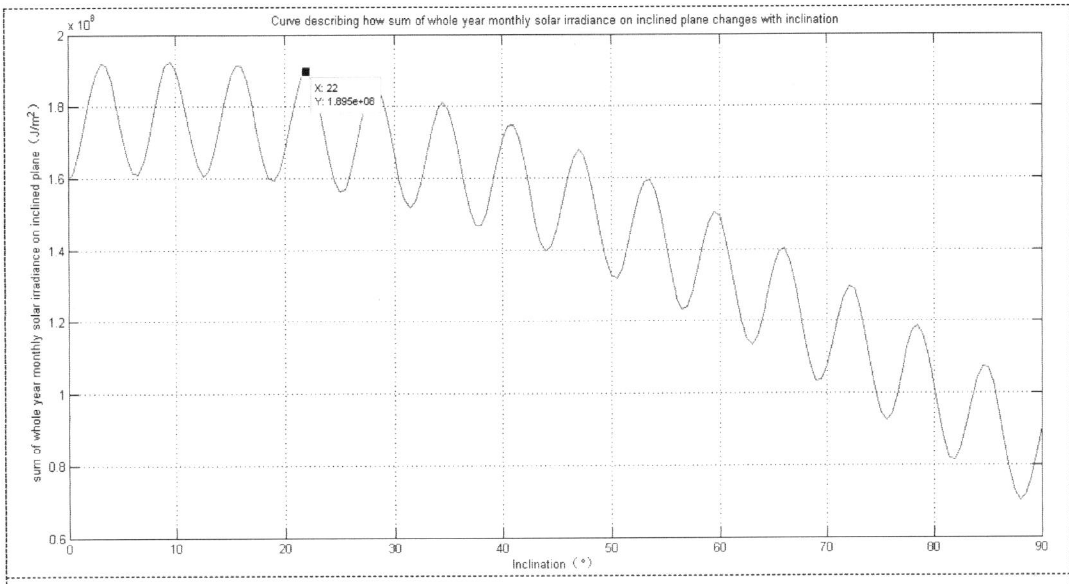

Figure 5. Slope of solar radiation of the whole year varying with inclination.

Therefore, combine situations in both winter and the whole year and through the modelling and optimization of this system in MATLAB, the inclination angle of photovoltaic cell array is modified to $\boldsymbol{\beta} = \mathbf{22°}$ in Haikou City.

5. Conclusions

In this paper, the practical problem about marine buoys has been solved. The design of ocean buoy's solar power system has been established through detailed calculation of solar radiation, modelling and configuration of photovoltaic modules and battery pack. This system can provide 1kW, 36V power for ocean buoys in South China Sea, and in extreme cases it can supply up to 72 hours of continuous power. The simulation system has been optimized and expanded to make the calculation more accurate. At the same time, the configuration of the solar power system can be realized in 28 other major cities nationwide.

References

[1] Jianbo Bai. Solar photovoltaic power generation system modelling, simulation and optimization [M]. Electronics Industry Press, 2014.

[2] PAN Jinjun, SHEN Yanbo, BI Zeqiang, WANG Xiangyun. Effect of Meteorological Factors on Solar Panel Temperature[J]. Journal of Applied Meteorology, 2014,25 (2).

[3] XUE Jiyuan, FENG Wenlin, ZHAO Fen, YANG Xiaozhan, Research on output characteristics and practical application of solar panels[J]. Infrared and Laser Engineering, 2015,44 (1).

[4] Klein, S.A. Comment on "The use of weighted Rb factors in calculating monthly-average isolation on titled surfaces" by J. R. Simonson[J]. Solar Energy, 1986, 36(6): 481-497.

[5] Klein, S.A. and J. C. Theilacker. Algorithm For Calculating Monthly-Average Radiation On Inclined Surfaces[J]. Journal of Solar Energy Engineering, Transactions of the ASME, 1981, 103(1): 29-33.

[6] Erbs, D.G,S.A.Klein and J.A.Duffie.Estimation of the diffuse radiation fraction for hourly, daily and monthly-average global radiation[J]. Solar Energy,1982,28(4):293-302.

Research on UAV Adaptive Landing Gear Control System

Dada Hu[a], Yang Li[b], Mingyang Xu[c] and Zhiyong Tang[d]

Beihang University, No.37 Xueyuan Road, Haidian District, Beijing, China

[a] 510587923@qq.com; [b] buaa12031206@126.com; [c] xmybuaa2015@126.com; [d] 57523523@qq.com

Abstract. Aiming at the severe requirement of landing terrain environment for UAVs, especially large UAVs, an adaptive landing gear is designed in this paper. The control system and PID controller based on attitude feedback, height feedback and force feedback are designed synthetically using high precision force sensor, attitude sensor and laser ranging sensor. The SIMULINK model is built in MATLAB and debugged in the actual system. The simulation and physical experiments show that the control system in this paper has the characteristics of good control robustness and fast response speed, which provides a beneficial attempt for engineering application of adaptive landing gear.

Keywords: self-adaptive landing gear, PID controller, attitude feedback, altitude feedback, force distribution

1. Introduction

UAV processed great potentials in both civil and military applications because of its excellent characteristics such as small volume, simple structure, flexible operation, vertical take-off and landing, etc.[1] During take-off and landing process, the height and velocity of an aircraft was very low, and the control ability was reduced.[2] However, the slope of the landing surface can severely limit viable landing areas.[3] As a result, take-off and landing process can be dangerous. When an aircraft experiences a hard landing, it must be inspected for damage before its next flight,[4] which cost a lot of time. An adaptive landing gear can solve the problem.

Some research institutions have made some progress in the development of adaptive landing gear. As figure1 shows, a recent example of robotic landing gear was developed by the Georgia Institute of Technology under the DARPA's Mission Adaptive Rotor (MAR) program consisting of four articulated robotic legs that adapt their position to ensure that the helicopter stays level during landing.[5] During the landing process, each leg can be adjusted in real time to ensure the helicopter remains stable and minimizes the risk of the propeller touching the ground.

Figure 1. DARPA Robotic Landing Gear.

This paper mainly completed the following work: mechanical design, electronic circuit design, control algorithm design and software implementation of adaptive landing gear. In the end, the SIMULINK system simulation was built to test whether the system achieves the desired goal.

2. Platform building

Mechanical platform is designed to carry 470 helicopters to accomplish take-off and landing missions stably in different terrains, which is built mainly by plastic with 3d printing technology as figure2 shows. In theory, the adaptive landing gear only needs two degrees of freedom: attitude control in the pitch and roll directions. However, for future function expansion, the landing gear platform is finally designed into four legs. The four legs are designed can only move vertically, which brings two advantages: the plantar pressure on the four legs are relatively even, resulting in upgrade of structural life, and the plantar pressure on the centre of gravity arm are constant, which brings convenience to torque solution. The transmission structure is designed into four-bar linkage. This paper got suitable length of each link with mathematical tools to ensure that legs can barely move horizontally.

Figure 2. Mechanical Design of Adaptive Landing Gear Platform.

For aspects of actuator, sensors, and controllers, this paper chooses servo motor with reducer as actuators, output current and torque of the motor can be controlled by sending PWM pulse to the motor driver. The vertical pressure is measured at the plantar with mount force sensors. This paper installed mpu6050 module on the platform to receive three-axis acceleration, angular velocity, and Euler angel information. Ultrasonic range finder HY-SRF05 or laser range finder LIDAR Lite V2 was set up at the bottom of the platform to measure the height of the body to the ground. This paper also chooses stm32f429 minimum system board as controller for its powerful computing resources and abundant peripherals.

3. Attitude Dynamics Analysis and Controller Design

The primary goal of adaptive landing gear is to ensure that the UAV lands smoothly when it takes off, that is required to control the body's pitch and roll angle, and set the control target to 0 °. In this paper, the posture of the body is measured by the attitude sensor module, and finally the attitude of the body is controlled by adjusting the supporting force provided by the four legs respectively.

Figure 3 defines the number of legs and coordinate direction of the system. From a top-down perspective, the serial numbers of the four legs are 1, 2, 3 and 4 in the clockwise direction respectively. The coordinate system is the ENU system: the front of the body is the Y axis while the right is the X axis. The positive direction is determined according to the right hand rule: the pitch angle θ is around the X axis, and the roll angle Φ is around the Y axis.

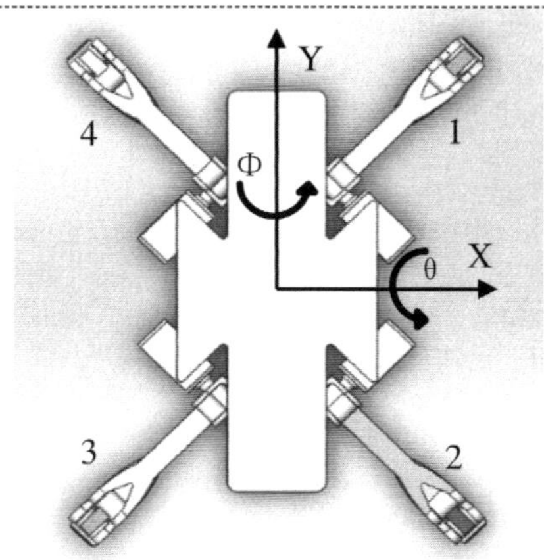

Figure 3. Coordinate system and definition of leg number.

The attitude dynamics equation is established as follows:

$$\begin{cases} \boldsymbol{J}\ddot{\boldsymbol{\Theta}} = \boldsymbol{T}_L + \boldsymbol{M}_G + \boldsymbol{T}_g \\ \dot{\boldsymbol{\Theta}} = \boldsymbol{\omega} \end{cases} \tag{1}$$

In formula 1, $\boldsymbol{J}$ is the moment of inertia of the body in the x-axis and the y-axis, $\boldsymbol{\Theta}$ is the Euler angle of the two axes, $\boldsymbol{\omega}$ is the angular velocity of the two axes, $\boldsymbol{T}_L$ is the torque of the four leg plantar pressure relative to the geometric center of the body, $\boldsymbol{M}_G$ is the gyro torque generated by Z-axis rotor's rotation on the x-axis and y-axis, $\boldsymbol{T}_g$ is the torque generated by deviation between centre of gravity and geometric centre. In this system, we tried to ensure the centre of gravity and geometric centre on the same vertical line, so the impact of $\boldsymbol{T}_g$ can be ignored. $\boldsymbol{M}_G$ expression is as follows:

$$\boldsymbol{M}_G = \boldsymbol{J}_R\boldsymbol{\omega}_R \times \boldsymbol{\omega} \tag{2}$$

In formula 2, $\boldsymbol{J}_R$ represents the moment of inertia of the rotor to the z-axis, and $\boldsymbol{\omega}_R$ represents the angular velocity at which the rotor rotates. For a UAV, the gyroscopic torque can be neglected due to the small value of $\boldsymbol{\omega}$. Therefore, the system attitude dynamics model is simplified as:

$$\boldsymbol{J}\ddot{\boldsymbol{\Theta}} = \boldsymbol{T}_L \tag{3}$$

The matrix form is:

$$\begin{bmatrix} J_x\ddot{\theta} \\ J_y\ddot{\Phi} \end{bmatrix} = \begin{bmatrix} T_1d_y - T_2d_y - T_3d_y + T_4d_y \\ -T_1d_x - T_2d_x + T_3d_x + T_4d_x \end{bmatrix} \tag{4}$$

J_x and J_y represent the moment of inertia of the frame to the x-axis and the y-axis, T_1 through T_4 respectively represent the plantar pressures on the four legs, d_x represents the distance from the plantar to the center of gravity in x axis, and d_y represents the distance from the plantar to the centre of gravity in y axis.

In order to improve the robustness of the system, attitude control system adopts the form of attitude controller as outer loop and angular velocity controller as inner loop. The goal of attitude controller is: after knowing target attitude angle $\boldsymbol{\Theta}_0 = [\theta_0 \;\; \Phi_0]$, controller $\boldsymbol{\omega}_a$ is designed to make $lim_{t\to\infty}\|\boldsymbol{e}_\Theta(t)\| = 0$,where $\boldsymbol{e}_\Theta = \boldsymbol{\Theta} - \boldsymbol{\Theta}_0$.For $\dot{\boldsymbol{\Theta}} = \boldsymbol{\omega}$, a PID controller is designed to determine the angular velocity expectation $\boldsymbol{\omega}_a$:

$$\boldsymbol{\omega}_a = -\boldsymbol{K}_{\Theta p}\boldsymbol{e}_\Theta - \boldsymbol{K}_{\Theta i}\int \boldsymbol{e}_\Theta - \boldsymbol{K}_{\Theta d}\dot{\boldsymbol{e}}_\Theta \tag{5}$$

The goal of angular velocity controller is: after knowing target angular velocity $\boldsymbol{\omega}_0$, controller $\boldsymbol{\tau}_a$ is designed to make $lim_{t\to\infty}\|\boldsymbol{e}_\omega(t)\| = 0$, where $\boldsymbol{e}_\omega = \boldsymbol{\omega} - \boldsymbol{\omega}_0$. For $\boldsymbol{J}\ddot{\boldsymbol{\Theta}} = \boldsymbol{T}_L$, a PID controller is designed to determine the torque expectation $\boldsymbol{\tau}_a$:

$$\boldsymbol{\tau}_a = -\boldsymbol{K}_{\omega p}\boldsymbol{e}_\omega - \boldsymbol{K}_{\omega i}\int \boldsymbol{e}_\omega - \boldsymbol{K}_{\omega d}\dot{\boldsymbol{e}}_\omega \tag{6}$$

4. Dynamics Analysis and Altitude Controller Design

The adaptive landing gear needs to ensure that the distance between the UAV's body and the ground is stable and the distance can be controlled according to the terrain condition or other instructions. In this paper, the body altitude data can obtain from laser range finder or ultrasonic range finder, and z-axis velocity data can obtain from the derivation of altitude data, or the integration of z-axis acceleration data. Finally, the total supporting force provided by the four legs is adjusted to control the altitude of the UAV.

So, the dynamics model is as follows:

$$\begin{cases} m\dot{v}_z = T_1 + T_2 + T_3 + T_4 - mg + f_z \\ \dot{z} = v_z \end{cases} \tag{7}$$

In formula 7, m is the total mass of the system, z is the distance between the body and the ground, v_z is the z-axis velocity of the body, f_z is the disturbance of z-axis, including the gravity of equipment mounted on UAV, elevating force provided by rotors and so on.

For the reason that f_z is uncontrollable, altitude control system adopts the form of altitude controller as outer loop and velocity controller as inner loop to ensure that the system perform well in different missions. The goal of altitude controller is: after knowing target altitude z_0, controller a_z is designed to make $lim_{t\to\infty}\|\Delta z(t)\| = 0$, where $\Delta z = z - z_0$. For $\dot{z} = v_z$, a PID controller is then designed to determine the z-axis velocity expectation v_z:

$$v_z = -K_{zp}\Delta z - K_{zi}\int \Delta z - K_{zd}\dot{\Delta z} \tag{8}$$

The goal of velocity controller is: after knowing target velocity v_0, controller T_z is designed to make $\lim_{t\to\infty}\|\Delta v(t)\| = 0$, where $\Delta v = v - v_0$. For $m\dot{v}_z = T_1 + T_2 + T_3 + T_4 - mg$, a PID controller is designed to determine the total supporting force expectation T_z:

$$T_z = -K_{vp}\Delta v - K_{vi}\int \Delta v - K_{vd}\dot{\Delta v} \tag{9}$$

5. Optimization of Landing Gear Force Distribution

In this paper, the system has four drivers to control four degrees of freedom, two degrees for attitude control, one degree for altitude control of the z-axis, and the last one can be used to distribute the forces on each leg and optimize working status of legs. $D(T)$ was set as the variance of four leg forces T:

$$\begin{cases} T = [T_1\ T_2\ T_3\ T_4] \\ D(T) = E(T^2) - [E(T)]^2 \end{cases} \tag{10}$$

In order to make the pressure of each leg relatively even, T should be fitted to make $D(T)$ the minimum value.

From the above we can see the constraints are as follows:

$$\begin{bmatrix} J_x\ddot{\theta} \\ J_y\ddot{\Phi} \\ m\ddot{z} + mg \end{bmatrix} = \begin{bmatrix} T_1d_y - T_2d_y - T_3d_y + T_4d_y \\ -T_1d_x - T_2d_x + T_3d_x + T_4d_x \\ T_1 + T_2 + T_3 + T_4 \end{bmatrix} \tag{11}$$

Formula11 can be simplified to the following form:

$$\begin{cases} T_1 - T_2 - T_3 + T_4 = X \\ -T_1 - T_2 + T_3 + T_4 = Y \\ T_1 + T_2 + T_3 + T_4 = Z \end{cases} \tag{12}$$

From Formula12 we get:

$$T_1 = \frac{X+Z}{2} - T_4,\ T_2 = -\frac{X+Y}{2} + T_4, T_3 = \frac{Y+Z}{2} - T_4 \tag{13}$$

Bring Formula13 in $D(T)$:

$$D(T) = E(T^2) - [E(T)]^2$$

$$D(T) = \frac{T_1{}^2 + T_2{}^2 + T_3{}^2 + T_4{}^2}{4} - \left(\frac{T_1 + T_2 + T_3 + T_4}{4}\right)^2$$

$$D(T) = \frac{T_1{}^2+T_2{}^2+T_3{}^2+T_4{}^2}{4} - \frac{Z}{16}^2 \tag{14}$$

From Formula14 we get if $T_1{}^2 + T_2{}^2 + T_3{}^2 + T_4{}^2$ take the minimum, $D(T)$ take the minimum, and:

$$T_1{}^2 + T_2{}^2 + T_3{}^2 + T_4{}^2 = 4T_4{}^2 - (2X + 2Y + 2Z)T_4 + \frac{(X+Z)^2+(X+Y)^2+(Y+Z)^2}{4} \tag{15}$$

If $T_4 = \frac{X+Y+Z}{4}$, $T_1{}^2 + T_2{}^2 + T_3{}^2 + T_4{}^2$ take the minimum, and $D(T)$ take the minimum, Bring T_4 in Formula13:

$$T_1 = \frac{X-Y+Z}{4}, T_2 = \frac{-X-Y+Z}{4}, T_3 = \frac{-X+Y+Z}{4}, T_4 = \frac{X+Y+Z}{4} \tag{16}$$

However, the leg force in this system is limited between 0N and 30N, which means that the input X, Y, Z need to be limited, otherwise leg force may be out of range. Solution to the following formula is:

$$0 \le T_1, T_2, T_3, T_4 \le 30 \tag{17}$$

We get:

$$|X| + |Y| \le min(120 - Z, Z) \tag{18}$$

The conclusion is that we only need to limit $|X| + |Y|$ value to ensure all leg force within a limited range. Therefore if $|X| + |Y| > min(120 - Z, Z)$, the values of $|X|$ and $|Y|$ are prorated:

$$X = \frac{X*min\,(120-Z,Z)}{|X|+|Y|}, Y = \frac{Y*min\,(120-Z,Z)}{|X|+|Y|} \tag{19}$$

6. Adaptive landing gear control system simulation

The decision-making process of adaptive landing gear system is as follows: when the drone is ready to take off, we can place the drone on the ground, start the attitude and altitude control channels, so that the UAV can adapt to any terrain until the drone successfully take off. When the drone works normally, reduce air resistance and prevent obstruction of the camera sight, stop the attitude and altitude control channels, and the landing gears all move to the highest position. When the drone switches to landing mode, the landing gear all moves down to the lowest position, and start leg force control channel, the force control target is set to 0. When the drone altitude drops to the set value $z_0(z_{min} < z_0 < z_{max})$, start the attitude and altitude control channels, the drone land on the ground stably, until the mission complete or restart. The decision flow chart is as Figure 4:

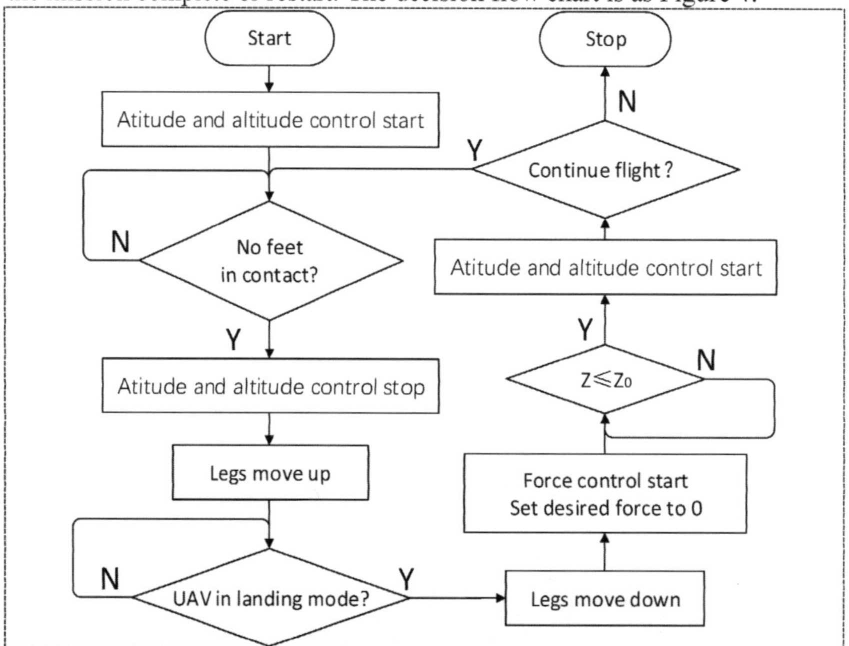

Figure 4. System decision flow chart.

IOP Conf. Series: Journal of Physics: Conf. Series **1061** (2018) 012019 doi:10.1088/1742-6596/1061/1/012019

Simulink simulation model of the system is shown in Figure 5:

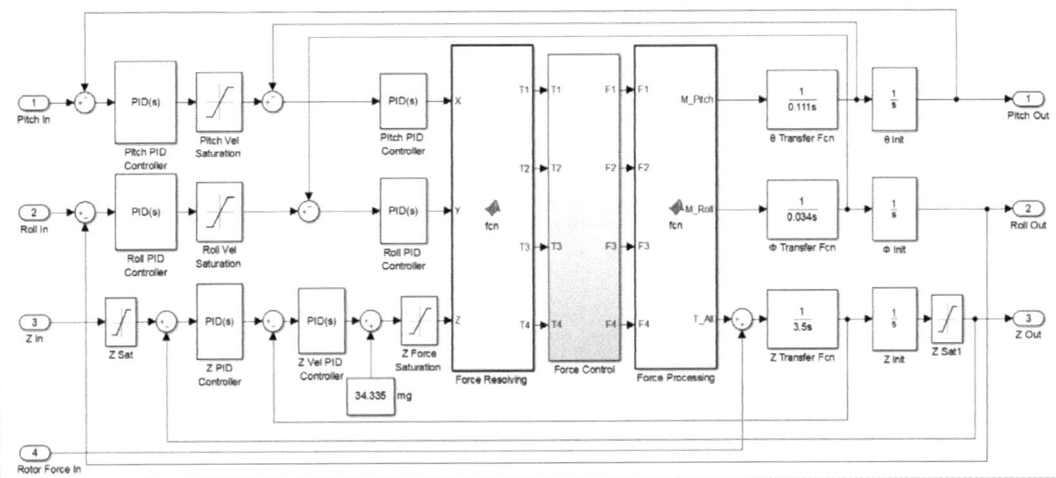

Figure 5. Simulation model of the system.

As shown in Figure 6, we set the initial pitch angle to 10 °, roll angle to -5 °, the current altitude to 0.2 meters. Target attitude angle are set to 0, target altitude to 0.2 meters, and the rotor elevating force are set to drop from the initial 30N to 0 slowly to simulate the actual landing conditions.

It can be seen from Figure 6 and 8 that there is a twist in attitude and altitude response, which is the result of extra lift force applied by the rotor. All leg forces are limited between 0 and 30N as is shown in Figure 9.

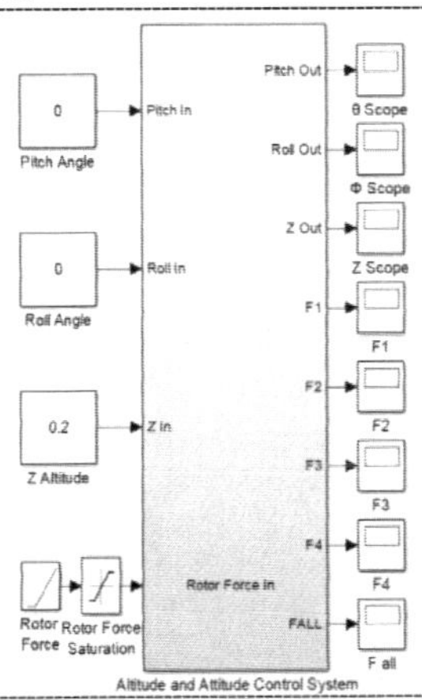

Figure 6. Condition set up.

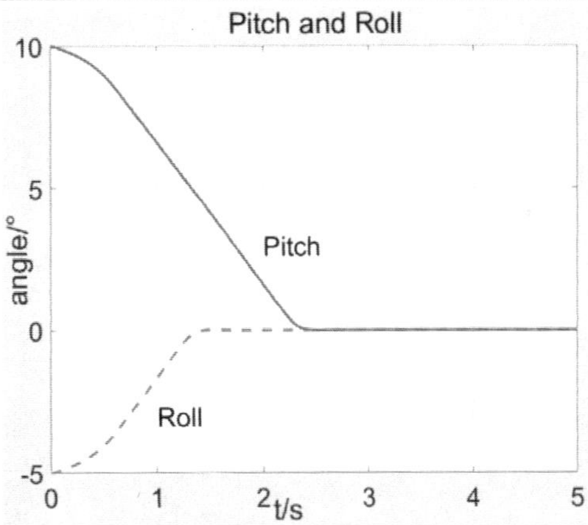

Figure 7. Attitude response.

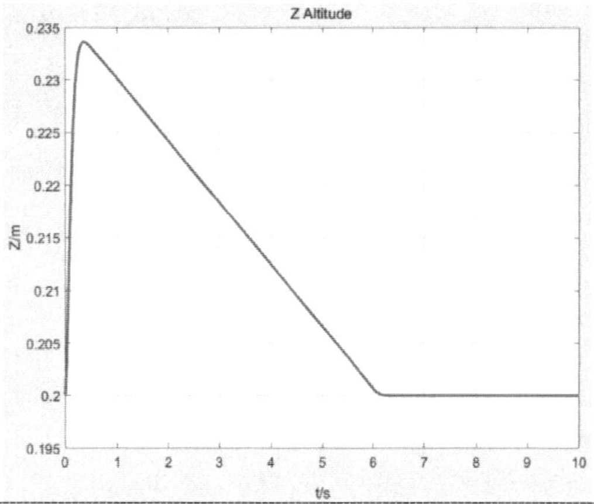

Figure 8. Altitude response.

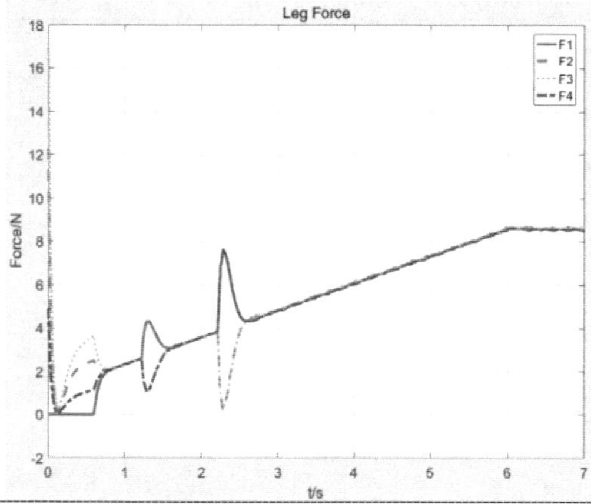

Figure 9. Leg forces.

7. Conclusions

In this paper, we completed the design of UAV landing gear, the construction of the platform, the algorithm design, the model and simulation of the system. Now the UAV can take off and land on most terrains. The simulation and experimental results show that the designed system has the advantages of simple structure, strong adaptive ability and good engineering application value.

References

[1] CP.Jiang. Research on Design and Realization of Control System for a Quadrotor Unmanned Aerial Vehicle [D]. Harbin Institute of Technology,2014.
[2] SL.Wang. Research on Design Technology of Automatic Take-off and Landing for UAV [D]. Nanjing University of Aeronautics and Astronautics,2011.
[3] Vasudevan Manivannan; Jared P. Langley; Mark F. Costello; Massimo Ruzzene. Rotorcraft Slope Landings with Articulated Landing Gear[D]. Georgia Institute of Technology, Atlanta, 2013.
[4] J. Kiefer; M. Ward; M. Costello. Rotorcraft Hard Landing Mitigation Using Robotic Landing Gear[D]. Journal of Dynamic Systems, Measurement, and Control, 2016.
[5] Daniel Melia Boix; Keng Goh; James McWhinnie. Modelling and Control of Helicopter Robotic Landing Gear for Uneven Ground Conditions [D]. School of Engineering and Built Environment, Edinburgh Napier University, Edinburgh, EH105DT, United Kingdom

IOP Conf. Series: Journal of Physics: Conf. Series **1061** (2018) 012020 doi:10.1088/1742-6596/1061/1/012020

Adaptive Positive-Negative Selection Approach

Piotr Hońko
Faculty of Computer Science, Bialystok University of Technology,
Wiejska 45A, 15-351 Białystok, Poland
E-mail: p.honko@pb.edu.pl

Abstract. Positive selection and negative selection are the main artificial immune approaches. They can find application in any task where an automatic identification of incoming cases as self or non-self is needed. Both the approaches build the so-called detectors for protecting self cells, i.e. positive class objects. In a positive selection approach, detectors are able to recognize self objects, whereas the other approach produces detectors that eliminate non-self objects. The difference in construction of detectors causes that the choice of the approach may depend on the characteristic of the data under consideration.

The goal of this paper is to provide a hybrid approach that is able to adapt to a given data set, producing thereby a classifier achieving the best performance. The adaptation is done twofold: (a) in the training phase, each candidate for a detector is automatically determined whether it will be used to build a self or non-self detector; (b) in the inner test phase, the best type of the classifier (i.e. pure positive selection, pure negative selection, positive-negative selection, negative-positive selection) is chosen.

Results of experiments conducted on databases from the UCI Repository show that the combination of positive and negative selection approaches may give a higher classification accuracy.

1. Introduction

The ability of a natural immune system to distinguish between pathogens and non-pathogens has inspired researchers to adapt this mechanism in artificial intelligence [1]. Such a system includes immune cells (detectors), called T cells, which have a component for recognizing self cells and an antigen receptor for locating and eliminating infected pathogens. The system generates a number of T cells, thereby any pathogen to infect the organism is very likely to be detected by at least one of cells. Before becoming a detector of the system a cell is verified to ensure that it is able to distinguish between self and non-self cells. The verification can be done using positive or negative selection. In positive selection T cell that recognizes any own cell is saved, otherwise it is removed. In negative selection, in turn, only cells that do not detect self cells are stored [2].

Both the approaches have successfully been adapted in the field of network security in order to automatically detect anomalies, intrusions, spams, or viruses (see e.g. [3, 4, 5, 6, 7, 8, 9, 10, 11]).

In spite of the fact that both selection approaches are mutually dual, thereby they can be complementary to each other, they have been usually used separately. Only a few studies on combining positive and negative selection approaches can be found.

A formal framework for analyzing different positive and negative detection schemes for data represented by binary strings was given in [12]. The framework enables to find the tradeoff between positive and negative detection schemes in terms of the number of detectors needed to correctly distinguish self objects from non-self ones.

An approach proposed in [13] is devoted to binary data representations and uses r-chunk matching rule. The approach starts with building binary trees corresponding to positive r-chunk detectors. To obtain a compact representation of the positive detector set, all complete subtrees of these trees are removed. Finally, a positive tree is replaced with its negative counterpart if the latter one is more compact.

The goal of this paper is to propose a combination of positive and negative selection approaches that adapts to the data to be classified. In the training phase, points from real-valued space (i.e. candidates for detectors) are selected randomly. Based on the distance of a point to the nearest self object it is determined whether it will be a self or non-self detector. In the inner test phase, objects are classified using four combinations: pure positive selection, pure negative selection, positive-negative selection, negative-positive selection. In the first two cases, only self or non-self detectors are used. In the third (fourth) case, if an object is not recognized by any self (non-self) detector, then it is classified by its nearest detector. One of the four combinations that gives the highest accuracy is chosen as the classifier for the data.

Experimental research shows that the positive-negative approach may increase the classification accuracy and that a better performance is obtained when one of the four combinations is defined individually for each dataset.

The rest of the paper is organized as follows. Section 2 proposes a combination of positive and negative selection approaches. Section 3 describes experimental research conducted using the proposed approach and six datasets taken from the UCI Repository. Concluding remarks are provided in Section 4.

2. A Combination of Positive and Negative Selection Approaches

This section proposes an algorithm (Algorithm 1) that generates self and non-self detectors at the same time. An algorithm (Algorithm 2) that classifies new objects according to one of four versions (positive selection, negative selection, and their two combinations) is also introduced.

Each point (i.e. m-ary vector of real numbers) to be used for building a detector in Algorithm 1 is selected randomly from the data space that is assumed to be normalized to the hypersphere. The selection is done from either the whole space, i.e. R_{SN} or the space of self objects, i.e. R_S. The choice of the space is done at random. This procedure enables to avoid an undesirable disproportion between self and non-self detectors sets in terms of their cardinalities. Namely, selecting points from the whole space might cause that the construction of a self detector is very unlikely (i.e. a candidate for a detector is usually too far from the nearest self object). On the other hand, when we limit the domain to the self objects space, the number of constructed non-self detectors might be very small (i.e. a candidate for a detector is usually too close to the nearest self object).

The radius of a self detector is constant and is defined by the user. The radius of a non-self detector is defined individually and equals to the distance of the detector to nearest self objects.

The algorithm stops searching for new detectors if each of k points selected in a row is within the scope of any detector already created.

Besides the pure positive or negative selection, Algorithm 2 is able to use its combination for objects that are not included in the sphere of any detector. The class of the object (that is not inside any detector already generated) is defined based on the type of the detector (self or non-self) that is the nearest to the object. The distance is measured between the object and the detector center.

IOP Conf. Series: Journal of Physics: Conf. Series **1061** (2018) 012020 doi:10.1088/1742-6596/1061/1/012020

Algorithm 1: *GenerateDetectors*

```
Data: S – a set of self cells; r – the self detector radius; k – the maximal number of
      failed attempts in a row;
Result: D_S, D_NS – sets of self and non-self detectors, resp.;
begin
    R_S := {[minS_{x_i}, maxS_{x_i}] : 1 ≤ i ≤ m};
    /* The set of ranges of variable domains from S                         */
    /* (S_{x_i} is the domain of a variable x_i);                            */
    R_NS := [0,1]^m;
    D_S = ∅; D_NS = ∅; k' = k;
    while k > 0 do
        b = getRandomValue({0,1});
        /* returns a random value from the set {0,1}                        */
        if b = 1 then R = R_S else R = R_NS;
        p = getRandomPoint(R);
        /* returns a random point from the space defined by R               */
        b := 1;
        if isInSideAnyDet(p, D_S) then b := 0;
        /* returns true if a point is inside any detector of a give set;    */
        else if isInsideAnyDet(p, D_NS) then b := 0;
        if b = 1 then
            v = distToNearestSelf(p, S);
            /* returns the distance of a point to the nearest self object;  */
            if v ≤ r then D_S = D_S ∪ {p};
            else D_NS = D_NS ∪ {(p, v)};
            k := k';
        end
        else k := k − 1;
    end
end
```

3. Experiments

This section describes experimental research that concerns verification of effectiveness of the proposed positive-negative selection approach.

Six datasets (Table 1) taken from the UCI Repository (http://archive.ics.uci.edu/ml) were used in the experimental research.

Table 1. Characteristics of datasets.

short name	full name	no. attr.	no. obj.	class distribution
Iris	Iris	4	150	50:50:50
Wine	Wine	13	178	59:71:48
Letters	Letter Recognition	16	20 000	789:19 211
Shuttle	Statlog (Shuttle)	10	43 500	34108:37:132:6748:2458:6:11
Skin	Skin Segmentation	3	245 057	50 857:194 200
Record	Record Linkage Comparison Patterns (10%)	12	574 913	2093:572 820

Algorithm 2: *ClassifyObjects*

```
Data: O – a set of objects; r – the self detector radius; D_s, D_ns – sets of self and
      non-self detectors, resp.; C – the combination (PS – positive selection, NS –
      negative selection, PNS – positive-negative selection, NPS – negative-positive
      selection);
Result: O_s, O_ns – sets of self and non-self objects, resp.;
begin
  foreach o ∈ O do
    if C ≠ NS then v = isInsideAnyDet_2(o, D_s);
    /* returns -1 if a point is inside any detector of a give set, or
       the distance to the nearest detector otherwise;                 */
    if C ≠ PS then v' = isInsideAnyDet_2(o, D_ns);
    switch C do
      case PS do
        if v = −1 then O_s := O_s ∪ {o}; else O_ns := O_ns ∪ {o};
      end
      case NS do
        if v' = −1 then O_ns := O_ns ∪ {o}; else O_s := O_s ∪ {o};
      end
      case PNS do
        if v = −1 then O_s := O_s ∪ {o};
        else if v ≤ v' then O_s := O_s ∪ {o}; else O_ns := O_ns ∪ {o};
      end
      case NPS do
        if v = −1 then O_ns := O_ns ∪ {o};
        else if v ≤ v' then O_s := O_s ∪ {o}; else O_ns := O_ns ∪ {o};
      end
    end
  end
end
```

The following testing procedure was applied. First of all, to avoid possible influence of different ranges of attribute domains, all database were normalized into a unitary hypercube $[0,1]^n$ where n is the cardinality of the attribute set. One class (the one whose cardinality is underlined in Table 1) of a database was taken as self data, the remaining ones as non-self data. Each data set was divided into five equinumerous sets. Four of them was used in 4-cross-validation (inner test), and the remaining one as an additional testing set (outer test). The non-self objects were only used in the classification phase. To eliminate the influence of randomness each data set was tested five times and the average results were taken. The radius of self objects (0.15 of the maximal possible distance) was defined in a preliminary experimentation. The distance between an object and detector was measured using the Euclidean metric. The quality measure used in the experiments was accuracy.

Table 2 shows the classification accuracy for the training data (TR), test data from 4-cross-validation ($TS1$), and additional test data ($TS2$). The negative-positive selection combination (NPS) had the best performance for five (TR), three ($TS1$), and three ($TS2$) out of six datasets. However, its results for other datasets are not always close to the best ones, see e.g. *Wine* ($TS1$, $TS2$) or *Record* (TR, $TS1$, $TS2$). It leads to the conclusion that the choice of the combination should be done for each database individually based on the training data (TR) result or first

Table 2. Accuracy of classification.

		Iris			*Wine*			*Letters*	
App.	TR	TS1	TS2	TR	TS1	TS2	TR	TS1	TS2
PS	99.93	99.00	99.30	97.51	**96.33**	**95.99**	67.41	67.29	67.16
NS	99.96	93.1	94.40	**99.89**	52.78	52.75	93.50	86.38	86.54
PNS	99.93	**99.60**	**99.90**	97.96	**96.33**	**95.99**	64.82	64.68	64.64
NPS	**100**	93.20	94.40	**99.89**	52.87	52.91	**93.77**	**86.66**	**86.78**
		Shuttle			*Skin*			*Record*	
App.	TR	TS1	TS2	TR	TS1	TS2	TR	TS1	TS2
PS	55.37	55.30	55.34	73.65	73.66	73.66	57.48	57.53	57.42
NS	**99.75**	**99.47**	**99.44**	95.74	95.73	95.74	**92.07**	**91.93**	**91.90**
PNS	56.15	56.13	56.12	72.35	72.36	72.36	70.97	71.00	70.93
NPS	**99.75**	**99.47**	**99.44**	**96.51**	**96.50**	**96.50**	70.97	71.00	70.93

test data ($TS1$) result. One can see that the best TR result translates into the best $TS1$ result for five data sets, whereas the highest $TS1$ result always implies the highest $TS2$ result. To investigate this issue more deeply, all the results, i.e. ones that were used to calculate the average results, were taken into account.

Table 3 shows in percentage how often the best result for TR ($TS1$) is also the best one for $TS1$ ($TS2$). The experimentation reveals that the choice of the combination is more accurate if it is done based on the test data ($TS1$). Namely, for five databases the result is better or the same, and the average performance is clearly higher.

Table 3. Convergence of the best results for the train and first test data and for the train and second test data.

	Iris	*Wine*	*Letters*	*Shuttle*	*Skin*	*Record*	*avg*
TR-> TS1	60.00	4.00	**100**	**100**	**100**	**96.00**	76.67
TS1->TS2	**81.33**	**100**	98.00	**100**	**100**	**96.00**	**95.89**

4. Conclusion

This paper has proposed a combination of positive and negative selection approaches devoted to real-valued data. The approach controls the process of detector generation, by determining whether a new detector will be self or non-self. This is also able to find the best combination (i.e. pure positive selection, pure negative selection, positive-negative selection, negative-positive selection) for a given dataset.

Based on the experimental research one can conclude that:

(i) A combination of both the approaches may give a higher classification accuracy than each of them used separately.

(ii) Finding the best combination for a given dataset is more likely when it is done based on not the training data but the test one.

The proposed approach is planned to be applied in future works as a tool for automatic detection of network anomalies.

Acknowledgments
This work was supported by the grant S/WI/1/2018 of the Polish Ministry of Science and Higher Education.

References

[1] Read M, Andrews P S and Timmis J 2012 *Handbook of Natural Computing* ed Rozenberg G, Bäck T and Kok J N (Berlin, Heidelberg: Springer Berlin Heidelberg) pp 1575–1597
[2] Dasgupta D 1999 *Artificial Immune Systems and Their Applications* ed Dasgupta D (Berlin, Heidelberg: Springer) pp 3–21
[3] Kim J and Bentley P J 2001 *Proceedings of the Genetic and Evolutionary Computation Conference (GECCO* (Morgan Kaufmann) pp 1330–1337
[4] Stibor T, Mohr P, Timmis J and Eckert C 2005 *Proceedings of the ACM SIGEVO Genetic and Evolutionary Computation Conference (GECCO-2005)* (Washington, D.C.: ACM Press)
[5] Stibor T, Timmis J and Eckert C 2005 *Proceedings of the Congress on Evolutionary Computation (CEC-2005)* (Edinburgh, UK: IEEE Press)
[6] Zhang F and Qi D 2011 *J. Comp. Vir.* **7** 267–277
[7] Chmielewski A and Wierzchon S T 2012 *Computer Information Systems and Industrial Management - 11th IFIP, TC 8 International Conference, CISIM 2012, Venice, Italy, September 26-28, 2012. Proceedings* pp 242–253
[8] Mostardinha P, Faria B F, Zúquete A and Vistulo de Abreu F 2012 *Artificial Immune Systems: 11th International Conference, ICARIS 2012, Taormina, Italy, August 28-31, 2012. Proceedings* ed Coello Coello C A, Greensmith J, Krasnogor N, Liò P, Nicosia G and Pavone M (Berlin, Heidelberg: Springer Berlin Heidelberg) pp 178–190
[9] Chikhi S and Ramdane C 2014 *Int. J. Inf. Sec. Priv.* **8** 1–25
[10] Peng L and Chen Y 2014 *2014 International Conference on Cyber-Enabled Distributed Computing and Knowledge Discovery, CyberC 2014, Shanghai, China, October 13-15, 2014* (IEEE Computer Society) pp 56–59
[11] Idris I and Selamat A 2014 *Appl. Soft Comput.* **22** 11–27
[12] Esponda F, Forrest S and Helman P 2004 *IEEE Transactions on Systems, Man, and Cybernetics, Part B (Cybernetics)* **34** 357–373
[13] Nguyen V T, Nguyen X H and Luong C M 2013 *The 2013 RIVF International Conference on Computing Communication Technologies - Research, Innovation, and Vision for Future (RIVF)* (IEEE Computer Society) pp 6–11

IOP Conf. Series: Journal of Physics: Conf. Series **1061** (2018) 012021 doi:10.1088/1742-6596/1061/1/012021

Tolerant *V*-Detector algorithm

Andrzej Chmielewski

Faculty of Computer Science, Białystok University of Technology, Wiejska 45a Street, 15-351 Białystok, Poland

E-mail: a.chmielewski@pb.edu.pl

Abstract. *V*-Detector algorithm is an example of negative selection algorithm, proposed to distinguish between *self* and *nonself* samples, represented as real-valued vectors. It is used mainly for detecting anomalies in high dimensional datasets. For this purpose, the set of detectors is generated, usually in random way, base only on information about normal behaviour. The main problem with this algorithm is scalability and high computational complexity. This paper presents a new approach for building the detectors, inspired by idea from the tolerance rough sets. With this algorithm it is possible to detect an outliers as well as those samples which are very similar to both classes and their correct classification requires additional verification. Partially, it also solves the scalability problem as detectors has a higher discrimination power in comparing to its basic version.

1. Introduction

The Negative Selection Algorithm (*NSA*) is inspired by the process of thymocytes (i.e. young T-lymphocytes) maturation in Nature Immune System, where only those lymphocytes survive which do not recognize any *self* molecule (own cell) [1]. A nice feature of NIS is that it does not need any example of pathogen (called *nonself*) to detect them. It means, such systems are able to detect even new and never met before harmful activities.

Many different *NSA* have been proposed over the last two decades (e.g. [2, 3, 4]). They have found wide range of applications in areas such as e.g. computer security, spam filtering, virus detection, handwritten character recognition (for more details, see, e.g. [5]). However, in many domains, there was encountered uncertain data and discrimination between *self* and *nonself* not always was possible. Some samples were very similar, regardless of used affinity (distance) metrics. They are very dangerous, especially in the security systems. In the case of uncertainty, they should be passed to additional post processing, to perform additional tests which allow for correct classification.

One of effective approaches for handling uncertainty in data is rough set theory [6, 7]. Some rough set inspirations have already been applied to NSA to deal with this problem. In [8] a lymphocytes, which are responsible for seeking and destroying the pathogens in NIS, were implemented as a decision rules, necessary to specify both: *self* and *nonself* classes. The proposed solution was implemented with rough valued lymphocytes to emulate approximate binding performed by NIS. In [9] rough sets were used for reducing the number attributes in AIS. In this case, the tests were conducted on KDD Cup 1999 high-dimensional dataset to detect anomalies in network connection.

In this paper, it is proposed the tolerant *V*-Detector algorithm, also inspired by RST. Similar approach was already presented in [10], where upper approximation, for both: binary and real-valued representations, were arbitrary selected base on the experimental results. Here, this value is dependent on number of the few nearest *self* samples.

2. Negative Selection Algorithm

The NSA, i.e. the negative selection algorithm, proposed by Forrest *et al.*, [1], is inspired by the process of thymocytes (i.e. young T-lymphocytes) maturation: only those lymphocytes survive which do not recognize any *self* molecules.

Formally, let $\mathcal{U}$ be a universe, i.e. the set of all possible molecules. The subset $\mathcal{S}$ of $\mathcal{U}$ represents the collection of all *self* molecules and its complement $\mathcal{N}$ in $\mathcal{U}$ represents all *nonself* molecules. Let $\mathfrak{D} \subset \mathcal{U}$ stand for a set of detectors and let $match(d, u)$ be a function (or a procedure) specifying if a detector $d \in \mathfrak{D}$ recognizes the molecule $u \in \mathcal{U}$. Usually, $match(d, u)$ is modelled by a distance metric or a similarity measure, i.e. we say that $match(d, u) = \texttt{true}$ only if $dist(d, u) \leq \delta$, where $dist$ is a distance and δ is a pre-specified threshold. Various matching function are discussed e.g. in [11] and [12].

The problem relies upon construction the set $\mathfrak{D}$ in such a way that

$$match(d, u) = \begin{cases} \texttt{false} & \text{if } u \in \mathcal{S} \\ \texttt{true} & \text{if } u \in \mathcal{N} \end{cases} \quad (1)$$

for any detector $d \in \mathfrak{D}$.

A naive solution to this problem, implied by the biological mechanism of negative selection, consists of five steps:

(a) initialize $\mathfrak{D}$ as empty set, $\mathfrak{D} = \emptyset$,
(b) generate randomly a detector d,
(c) if $match(d, s) = \texttt{false}$ for all $s \in \mathcal{S}$, add d to the set $\mathfrak{D}$,
(d) repeat steps (b) and (c) until the sufficient number of detectors is be generated.

3. V-Detector algorithm

The original V-Detector algorithm was proposed by Ji and Dasgupta in [3]. It operates on normalized vectors of real-valued attributes. Each vector can be viewed as a point in the d-dimensional unit hypercube, $\mathcal{U} = [0, 1]^d$. Each *self* sample, $s_i \in \mathcal{S}$, is represented as a hypersphere $s_i = (c_i, r_s)$, $i = 1, \ldots, l$, where l is the number of *self* samples, $c_i \in \mathcal{U}$ is the center of s_i and r_s is its radius. It is assumed that r_s is identical for all s_i's. Each point $u \in \mathcal{U}$ inside any *self* hypersphere s_i is considered as a *self* element.

For Euclidean affinity metric, the detectors d_j are represented as hyperspheres: $d_j = (c_j, r_j)$, $j = 1, \ldots, p$, where p is the number of detectors. In contrast to *self* elements, the radius r_j is not fixed but it is computed as the Euclidean distance from a randomly chosen center c_j to the nearest *self* element (this distance must be greater than r_s, otherwise the detector is not created). Formally, we define r_j as

$$r_j = \min_{1 \leq i \leq l} dist(c_j, c_i) - r_s. \quad (2)$$

In its original version, the V-Detector algorithm employs Euclidean distance to measure proximity between a pair of samples. Therefore, *self* samples and the detectors are hyperspheres. Formally, Euclidean distance is a special case of Minkowski norm L_m defined as:

$$L_m(\mathrm{x}, \mathrm{y}) = \left(\sum_{i=1}^{d} |x_i - y_i|^m \right)^{\frac{1}{m}}, \quad (3)$$

where x $= (x_1, x_2, \ldots, x_d)$ and y $= (y_1, y_2, \ldots, y_d)$ are points in $\Re^d$.

Particularly, L_2-norm is Euclidean distance, L_1-norm is Manhattan distance and L_∞ is Tchebyshev distance. Additionally, it was defined *fractional distance metric* (with $0 < m < 1$), which is more appropriate for high-dimensional spaces.

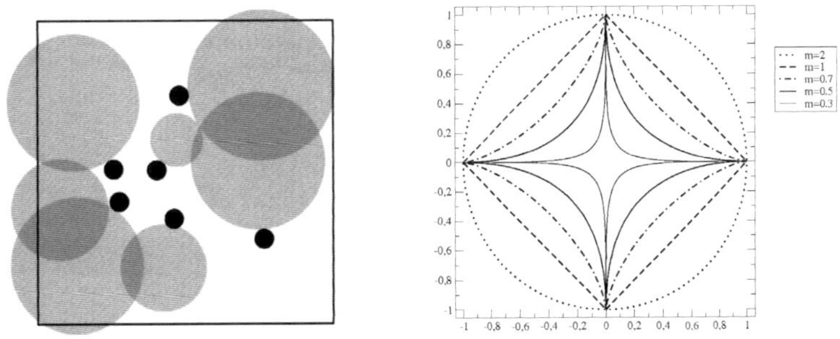

(a) (b)

Figure 1. (a) Example of performance V-Detector algorithm for 2-dimensional problem. Black and grey circles denotes *self* samples and v-detectors, respectively. (b) Unit spheres for selected L_m norms in 2D.

Another consequence of applying fractional metrics for V-Detector algorithm is modification of the shape of detectors. Figure 1(b) presents the unit spheres for selected L_m-norms in 2D with $m = 2$ (outer most), $1, 0.7, 0.5, 0.3$ (inner most).

4. Tolerant V-Detector

V-Detector algorithm is sensitive on *self* samples which are outliers. They obstruct the $\mathcal{N}$ space coverage and make a larger number of detectors necessary. Tolerant V-Detector introduces an additional radius of tolerance r^t_j ($r^t_j >= r_j$), which is a distance to the nearest k-*self* sample, instead of the nearest one. Hence, each tolerant v-detector (v^t_k-detector) in $\Re^d$ space is represented as a real-valued vector $d^t_j = (c_j, r_j, r^t_j)$.

Introducing the two different radii for a single detector d_j was inspired by RST, where r_j is responsible for the classification of all *nonself* samples with full certainty (lower approximation), while r^t_j (upper approximation) is used for determining objects which may possibly be *nonself*. Base on this assumptions, during the classification stage, censored sample is classified as:

- *nonself*- when was covered by at least one hypersphere with lower approximation r_j;
- *self*- when was not covered by any hypersphere with upper approximation r^t_j;
- *uncertain* - otherwise (when was covered by at least one hypersphere with upper approximation r^t_j.

Due to during the learning stage, the distance to all samples have to be calculated, r^t_j can be determined without increasing the time and computational complexity, in comparing to original V -Detector algorithm. While the censoring stage, additional verification should be performed only for not recognized samples. However, this operation does not have to be done immediately, since it only applies t o uncertainty objects (for example s uspicious network connections) and can be done in t he i dle t ime of t he processor. Figure 2 presents an example of vtk-detector in 2D space for k = 2 and Euclidean affinity metric.

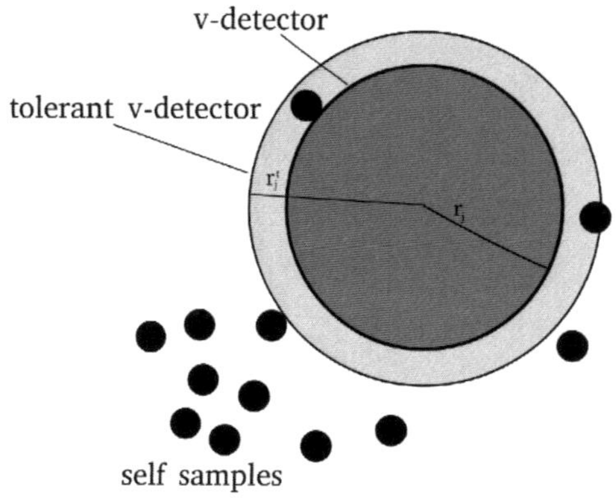

Figure 2. Example of v_k^t-detector in 2D space for $k = 2$ and Euclidean affinity metric.

5. Experiments

Experiments were performed on two very popular datasets, mainly used for testing intrusion detection systems, namely: KDD Cup 1999 and Kyoto 2006+. Both of them are high dimensional and numerous. To decrease the computation complexity, all tests were performed on samples with specified protocol: ICMP, TCP and UDP. In this paper, only results for ICMP are presented and detailed information about used subsets are presented in Table 1.

Table 1. Description of testing datasets.

Dataset	Protocol	No. of attributes	No. of samples	No. of *self* samples
KDD Cup 1999	ICMP	22	11911	4428
Kyoto+	ICMP	21	4069	319

Moreover, from the both subsets, some attributes were excluded (for example, source and destination MAC addresses in case Kyoto+) due to a met problems with normalization (too many unique symbolic values). In case of KDD Cup 1999, all attributes not related with ICMP protocol were removed. Was obtained a subset consisting of 22 attributes, instead of the original 42.

Tables 2 and 3 shows the obtained results for Kyoto+ and KDD 1999 Cup, respectively. For both testing datasets, it can be observed an increase in length of self radius (Δr) with increasing the value k. As a result, each detector were able to detect more *nonself* samples at the expense of FAR. However, the influence of k value on increase of DR for KDD 1999 Cup was more significant. For $k = 30$, DR has increased, more than twice, up to 78% from basic 33.9% at the very low FAR (less than 5%). For Kyoto+, for the same k value, DR has increased from 12% to 93%. However, in this case, over 41% *self* sample were not correctly classified. Hence, the lower k values should be chosen to achieve good balance between DR and FAR.

The column $|V_t|$ contains the percentage of samples classified as *uncertain*. As you might expect, this value increases with k up to 80% for $k = 30$. Hence, k should be carefully selected to minimize the number of samples, which further should be additionally censored by other algorithms.

The results presented in Tables 2 and 3 were achieved for $L_{0.5}$ affinity metric because dimensionality of testing datasets were to high for Euclidean and Manhattan distances.

IOP Conf. Series: Journal of Physics: Conf. Series **1061** (2018) 012021 doi:10.1088/1742-6596/1061/1/012021

Table 2. Results for Kyoto+ dataset.

k	Δr [%]	DR[%]	FAR[%]	$\|V_t\|$ [%]
0	-	12.8	-	-
1	11.8	30.9	4.2	16.1
2	17.3	39.7	7.9	24.2
3	19.3	43.5	8.8	26.6
5	26.5	59.8	13.9	45.1
10	36.9	73.0	23.8	54.3
20	51.6	87.4	34.1	67.8
30	58.2	93.6	41.5	73.6

Table 3. Results for KDD 1999 Cup dataset.

k	Δr [%]	DR[%]	FAR[%]	$\|V_t\|$ [%]
0	-	33.84	-	-
1	3.2	48.2	0.2	6.1
2	6.4	61.5	0.4	16.6
3	9.1	48.9	0.6	32.2
5	10.0	58.0	0.8	34.1
10	18.1	59.7	1.2	34.9
20	18.5	77.9	3.7	42.3
30	21.5	78.2	4.7	80.0

6. Conclusions

Application of rough sets theory (RST) to the *NSA* allow for determining the lower and upper approximation spaces in $\mathcal{U}$. In this way, an additional class of objects can be defined, except the standard $\mathcal{S}$ and $\mathcal{N}$, used by default in the immune systems. The *uncertain* class, defined as a bounding region of rough sets, is used for storing the samples for which the affinity metric to both classes is very similar. For correct classification, an additional algorithms could be applied and then the overall detection rate can significantly increase. This may be particularly important e.g. for security systems where slight deviations from normal connections may be unnoticed. Application of RST is one of the way to deal with this problem.

It is worth to emphasizing, high ratio of DR was obtained without significant increasing the computational complexity. While the learning stage, the process of generating detectors does not require any additional operations as distances to all samples have to be calculated even in the basic version of V-Detector algorithm. During the censoring samples, only those not recognized are additionally compared with the radius representing the upper approximation.

Future works should focus on applying the RST to b-v model [4] as well as to other popular affinity metrics.

Acknowledgment

This research was partially supported by the grant S/WI/1/2018 of the Polish Ministry of Science and Higher Education.

References

[1] Forrest S., Perelson A., Allen L., Cherukuri R.: Self-nonself discrimination in a computer. In Proc. of the IEEE Symposium on Research in Security and Privacy, Los Alamitos, 1994, pp. 202–212.
[2] Forrest S., Hofmeyr S. A., Somayaji A., Longstaff T. A.: A sense of Self for Unix Processes. Proc. of the 1996 IEEE Symposium on Research in Security and Privacy, IEEE Computer Society Press, 1996, pp. 120–128.
[3] Ji Z., Dasgupta D.: Real-valued negative selection algorithm with variable-sized detectors. Genetic and Evolutionary Computation GECCO-2004, Part I, LNCS Vol. 3102, Seattle, WA, USA, Springer-Verlag, 2004, pp. 287–298.
[4] Chmielewski A., Wierzchoń S. T.: Hybrid negative selection approach for anomaly detection. In: A. Cortesi et al. (Eds.) Computer Information Systems and Industrial Management, LNCS 7564, Springer, 2012, pp. 242-253.
[5] Fernandes D.A.B., Freire M.M., Fazendeiro P.A.P., Incio P.R.M.: Applications of Artificial Immune Systems to Computer Security: A Survey. Journal of Information Security and Applications, Volume 35, August 2017, pp. 138–159.
[6] Pawlak Z.: Rough Sets. Theoretical Aspects of Reasoning about Data. Kluwer Academic, Dordrecht, 1991.
[7] Hońko P., Compound approximation spaces for relational data. International Journal of Approximate Reasoning Volume 71, April 2016, pp. 89–111.
[8] Félix R., Ushio T.: Rough Lymphocytes for Approximate Binding in Artificial Immune Systems. 15th International Conference on Electronics, Communications and Computers (CONIELECOMP 2005), Puebla, Mexico, pp. 272–277.
[9] Shen J., Wang J., Ai H.: An Improved Artificial Immune System-Based Network Intrusion Detection by Using Rough Set. Communications and Network, 2012, 4, pp. 41–47.
[10] Chmielewski A. (2017) Application of Rough Sets to Negative Selection Algorithms. In: Dang T., Wagner R., Kng J., Thoai N., Takizawa M., Neuhold E. (eds) Future Data and Security Engineering. FDSE 2017. Lecture Notes in Computer Science, vol 10646. Springer, Cham.
[11] Harmer P. K., Wiliams P. D., Gunsch G. H., Lamont G. B.: Artificial immune system architecture for computer security applications. IEEE Trans. on Evolutionary Computation, Vol. 6, 2002, pp. 252–280.
[12] Ji Z., Dasgupta D.: Revisiting negative selection algorithms. Evolutionary Computation, vol. 15(2), 2007, 223-251.

IOP Conf. Series: Journal of Physics: Conf. Series **1061** (2018) 012022 doi:10.1088/1742-6596/1061/1/012022

The Demand Response and UPFC Coordinated Frequency Control Strategy

Yang Yao[1,*], Xue-Li Chen[1], Si Zhang[2] and Shun-xiang Ji[1]

[1]Nanjing Normal University, Nanjing, China
[2]Nanjing Forestry University, Nanjing, China

*E-mail: 1183053846@qq.com

Abstract. In order to improve the stability of the power system, suppressing the low frequency oscillation of power system is already a problem that every country is eager to solve. Many power workers have done a lot of theoretical research and experimental verification on low-frequency oscillation and have got remarkable achievements. Energy from the Demand Side has been widely used in our daily life, and thousands of demand response loads have been used in frequency control whose controlling effects are quite obvious; meanwhile, because of the development of electrical technology, many smart grid devices can be fully used, and they have improved the stability of power system, which is consistent with the original intention of putting demand response loads into power system. UPFC can improve the system damping, and effectively suppress the low-frequency oscillation of the system as well as improve the stability of the power angle, so if UPFC and demand response are combined, the low frequency oscillation of the power system can be better suppressed.

1. Introduction

Demand response can provide added damping in oscillation control without bringing in any energy consumption, so it is environment and economy-friendly and more consistent with the socialist modernization construction; at the same time, with some time prolonged processes being eliminated, demand response control can be easier and sound than traditional control measures. In the stable aspect of power system, the Unified Power Flow Controller (UPFC) can achieve fast efficient adjustment to current, rational control to active and reactive power flow, improvement of circuit delivery capacity and optimization of the overall operation; in the dynamic aspect, it can dynamically support the voltage of the access points, and improve the general voltage stability through quick reactive compensation technology.

2. Controlling Methods and Mechanism of Demand Response

As the research system will be affected by small disturbances, the variation of variables can be seen as putting a small deviation to the original process. The variables can be:

$$\begin{aligned} &T'_{d0}\frac{d\Delta E'_q}{dt} = E_{qe} - [\Delta E'_q + \Delta I_d(x_d - x'_d)] \\ &\frac{d\Delta\delta}{dt} = \Delta w w_0 \\ &Tj\frac{d\Delta w}{dt} = \Delta P_T - \Delta P_e - D\Delta w \end{aligned} \tag{1}$$

The above equation should be the linear differential equation of synchronous generator.

Demand Response loads engaged in the low frequency oscillation of power system can be separated into two categories according to their characteristics. The first category is equipment that can not only change the state of switches but also continually adjust the amount of loads engaged in the power system, and the other category can only change the state of switches[1].

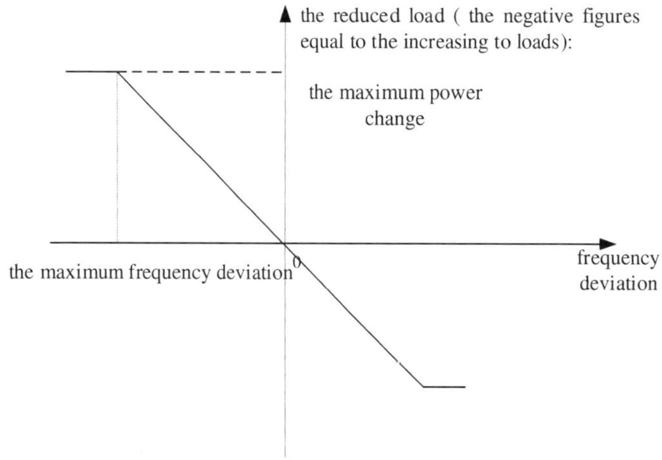

Figure 1. Frequency regulation characteristics of demand response loads

When a controller detects that the deviation Δf is lower than Δfth, the controller will disconnect the load so as to restore the electric frequency to a stable level. In order to prevent the negative impact of the instant disconnection of many loads and stagger the turn-off time, Δfth should be set to every engaged load. The current domestic and overseas solutions are mostly adopt instant measures to the Δfth of every equipment. While because of the Δfth of different equipment may influence the controlling effect directly, it is obvious that such measures are unsuitable[2].

Taking the frequency deviation caused by the active power deficiency as example, the amount of PDR that needs to be closed should be set as:

$$P_{DR} = \begin{cases} -P_{DRm} & \text{If } \Delta f > \Delta f_{\max} \\ -k_{DR} . \Delta f & \text{If } -\Delta f_{\max} \le \Delta f \le \Delta f_{\max} \\ P_{DRm} & \text{If } \Delta f < -\Delta f_{\max} \end{cases} \tag{2}$$

$\Delta f_{\max}$ is the maximum frequency deviation provided by engaged demand response loads; P_{DRm} is the total load engaged in the controlling system, and the coefficient k_{DR} should be:

$$k_{DR} = -\frac{P_{DRm}}{\Delta f_{prm}} \tag{3}$$

3. Design for Coordination Control of Demand Response and UPFC

The function of auxiliary controller is to increase system damping and suppress low-frequency oscillations by improving the control strategy of UPFC. Based on the damping characteristics introduced in the last chapter and the principle of power system stabilizer, the UPFC auxiliary controller shown in the Figure 1 is designed. Taking the line current I as reference direction, the voltage of series side is decomposed into 2 vectors Vp and Vq, the former one is parallel to I and the latter one is vertical to I. Then the active power and reactive power of the series side can be expressed as IVp and IVq [3], which are adjusted by changing Vp and Vq to carry out of line power. T_{w} in Figure 2 is a DC blocking time constant, T1 and T3 are time constants of lead compensation, and T2 and T4 are constants of lag compensation. This auxiliary controller adopts proportional control, taking the oscillation frequency of system Δω as input. It outputs the control signal after passing the DC blocking section and amplitude limiting unit, which is similar to the control of PSS [4].

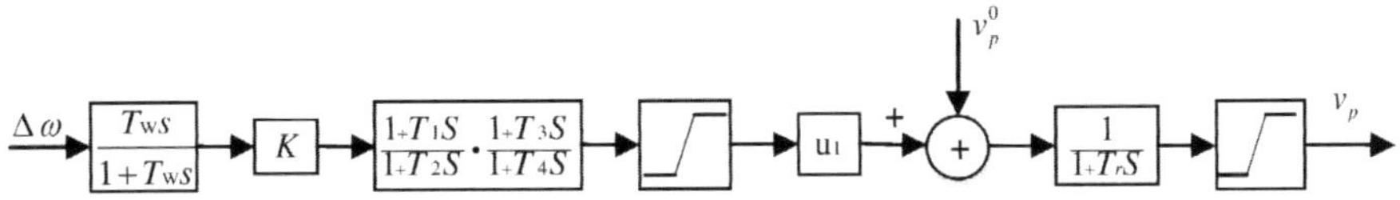

Figure 2. UPFC auxiliary controller block

Vp: This variable is the vector of the series voltage Vs, which is in phase with line current. In a steady state, input Vp0 is set as zero, making the active power interchange between UPFC and alternating current system only occurs when this variable is regulated by POD controller (during the transient period)[5].

4. Analysis of Examples

4.1. Single-machine infinite-bus system

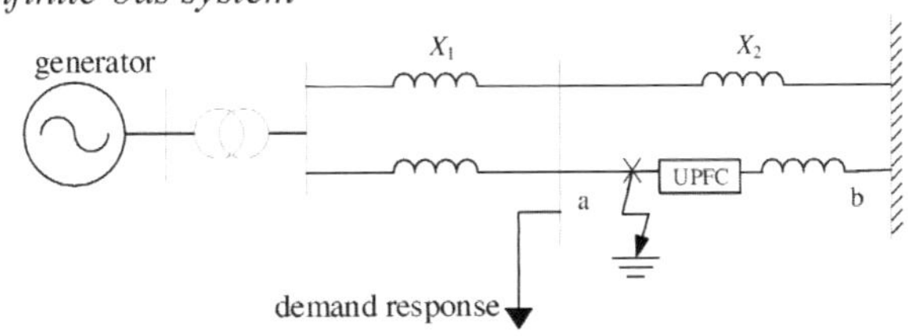

Figure 3. SMIB system

As the Figure 3 shows, what is different from the single-machine infinite-bus system model is that the load ab is equipped with UPFC, and UPFC is combined with demand response to suppress the low-frequency of the electrical power system. To increase additional damping, parameters including PDR, the amplification factor Kw and constants of compensation time T1 and T3 should be optimized (set T2 = T4 =0.05s) [6].

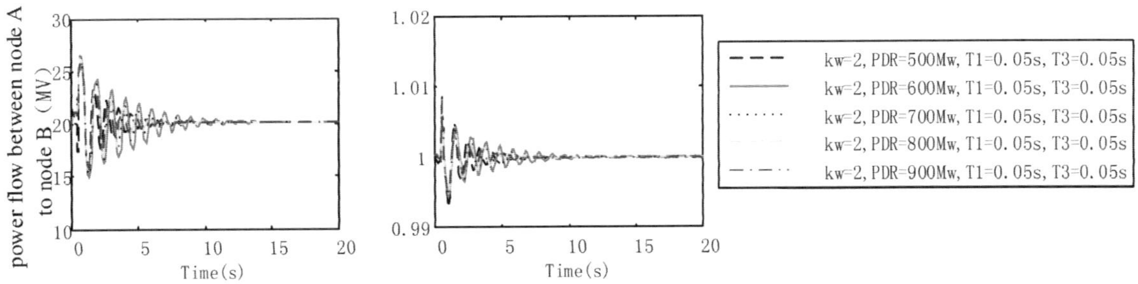

Figure 4. Simulation results of demand response with different capacities

Figure 4 shows the system eigenvalues with different numbers of demand response resources in the single-machine infinite system, and it suggests that higher quantities of demand response resources can provide more damping for the system.

In the following part, it will illustrate the influence that different values of the gain parameter Kw have on the low-frequency oscillation suppression effects in the electrical power system.

From the Figure 5, it can be inferred that, the value of Kw can affect the suppression effect to a great extent, when the demand response reaches a certain size (30% of the total load). In the process of the five experiments, when Kw is 0.6, the suppression effect is the best and the suppression of the low-frequency oscillation in power system takes the shortest time.

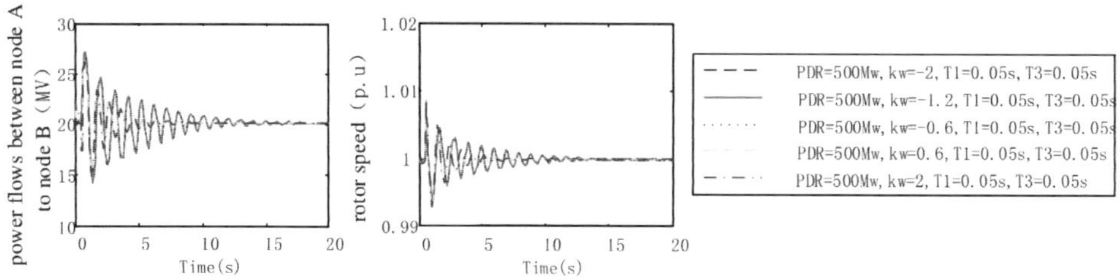

Figure 5. Simulation results under the conditions of different gains

The following part will discuss the suppression effect that the time advance constant T1 and the time delay constant T3 bring to the low-frequency oscillation in power system [7].

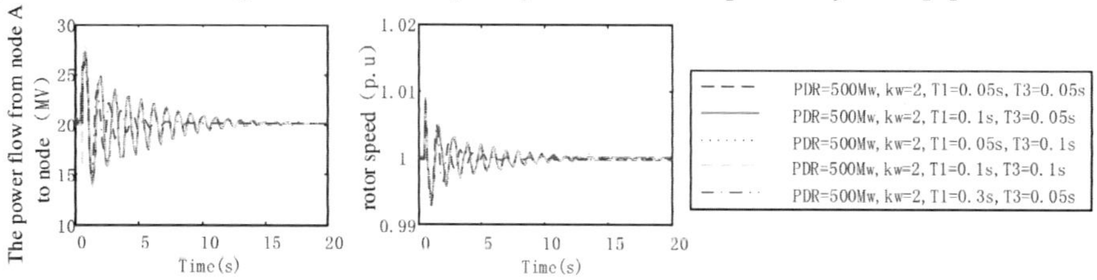

Figure 6. Simulation results under the conditions of different time constants

It can be inferred that, from Figure 6, different time parameters cause the difference of suppression effects, which is not very obvious. The additional damping is the most obvious only when T1 and T3 are about 0.5s.

Although many sets of data should be measured to optimize parameters, this paper does not list every set of data but adopts the three-dimensional Figure 7 to show the relationship between different Kw values and damping ratios because of the limited space.

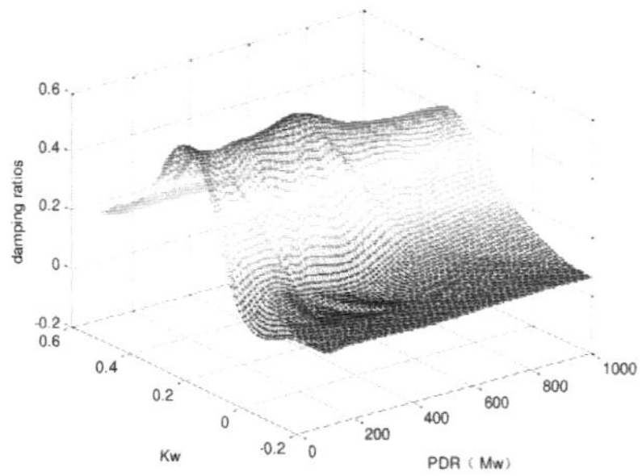

Figure 7. The relationship between different Kw values and damping ratios

According to the data mentioned above, set the objective function and get the PDR varying in the range of 200MW-1000MW, UPFC operation gain parameter Kw varying in the range of -3-3 and time parameters T1 and T3 varying in the range of 0.01s – 0.1s. This objective function, which is in the single-machine infinite bus system, reflects the coordination of demand response and UPFC to suppress low-frequency oscillation effect. The control variables are demand response, the gain parameter of UPFC controller Kw, time advance constant T1 and time delay constant T3.

The objective function: min $f=\lambda_1\sigma+\lambda_2\xi$.

After 100 iterations, the result is that f_val=-0.3243, when Kw=0.3846, PDR=645.6097kw, T1=0.0562s and T3=0.0643s.

4.2. The Two-area and Four-machine Power System

According to the report about LFC at home and abroad, the LFO of the tie line between two interconnected power systems with the lowest frequency of about 0.2~0.7Hz is called inter-area low frequency oscillation; the LFO between different power plants in the same area with the frequency of

IOP Conf. Series: Journal of Physics: Conf. Series **1061** (2018) 012022 doi:10.1088/1742-6596/1061/1/012022

about 1Hz is called regional low frequency oscillation; the LFO between different units in the same power plant with the highest frequency of about 1.5~2.5Hz is called in-plant low frequency oscillation[8].The Two-area and Four-machine Power System is shown in Figure 8.

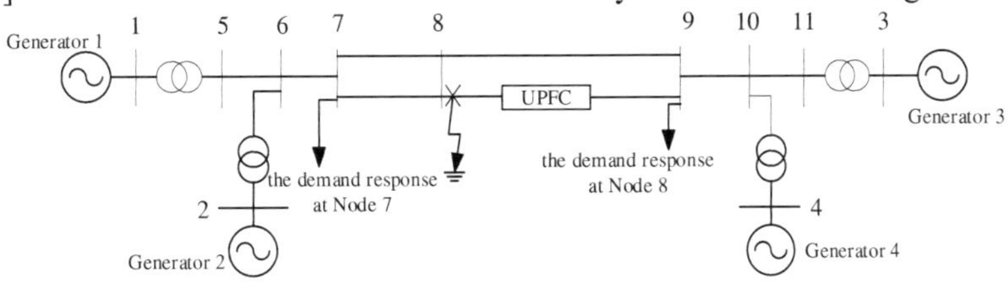

Figure 8. The Two-area and Four-machine Power System Model Engaged in Low Frequency Oscillation Suppression with UPFC

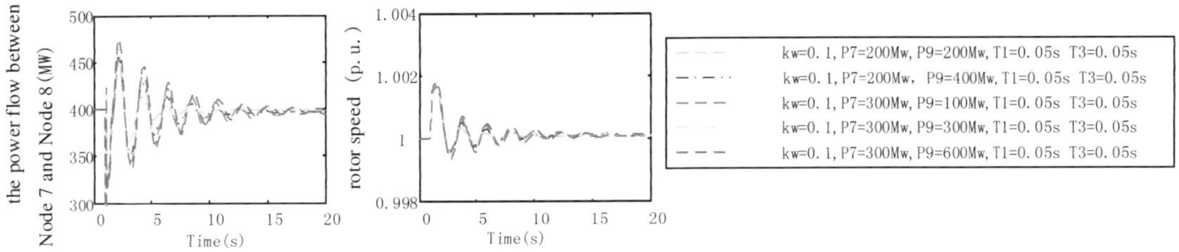

Figure 9. The Simulation Results under Demand Response with Different Capacities

Figure 9 shows the single machine infinite bus system eigenvalues under different numbers of available resources of demand response in the two-area and four-machine power system. It shows that a high quantity of demand response resources can be used to provide more damping for the system, and the demand response of the same quantity can produce different suppression effects because of the different locations of the two nodes. Next, the effect of various gain parameters Kw on the low frequency oscillation suppression under the two-area and four-machine power system will be discussed.

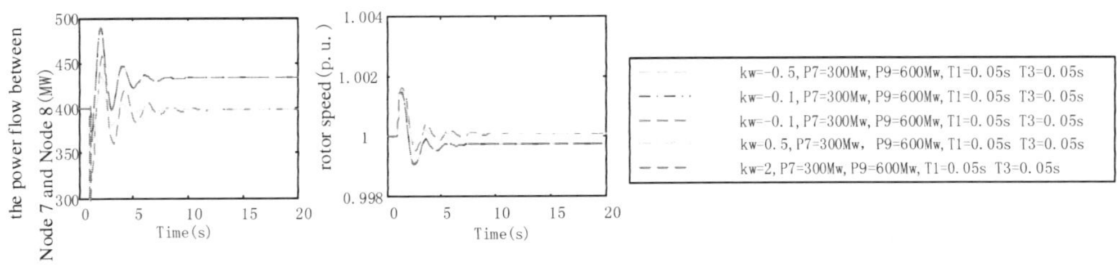

Figure 10. The Simulation Results under Different Gains

Figure 10 shows that if the demand response reaches a specific size (22% of the total load), the change of Kw value will largely influences the suppression effect. In the 5 experiments, when the Kw value is -0.5, the suppression effect is the best and the time of low frequency oscillation suppression of the power system is the shortest. When the Kw value is over 0.5, the real part of eigenvalue of equation of motion in the electronic system is positive and the damping ratio is negative, so the system is no longer stable.

Then, the suppression effect of the values of time-advance parameter T1 and time–delay parameter T3 on the low frequency oscillation of the power system will be discussed.

Figure 11 shows that different time parameters will affect the suppression effect. But the change is not obvious and only when the values of T1 and T3 are around 0.05s, the additional damping is the most obvious.

The parameters should be optimized, so it is necessary to measure and obtain multiple sets of data. For lack of space, the data will not be listed. Figure 12 is the three-dimensional Fig about the relation among damping ratio, different Kw values and PDR.

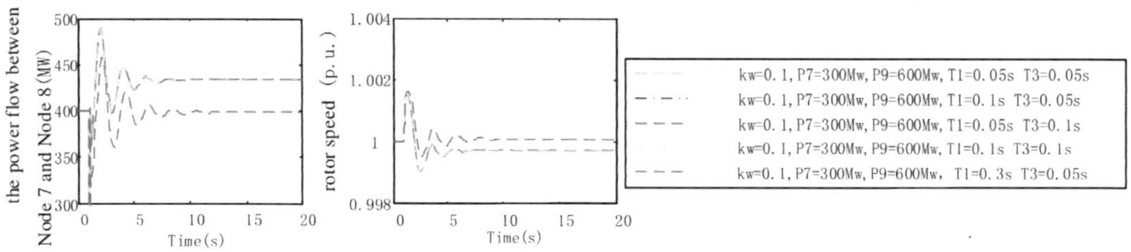

Figure 11. Simulation Results with Different Time Parameters

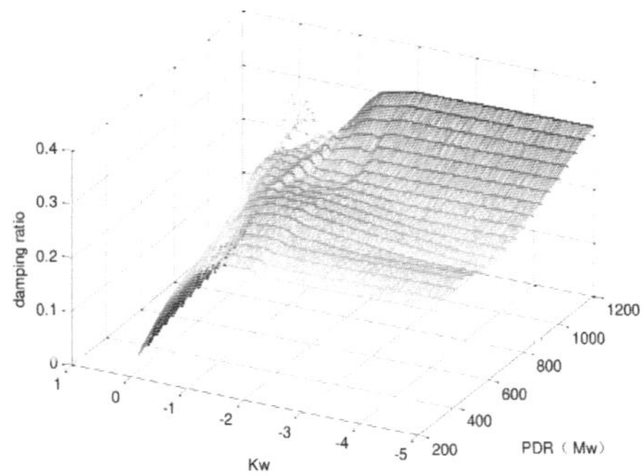

Figure 12. The relationship between different Kw values and damping ratios

The objective function: $\min f = \lambda_1\sigma + \lambda_2\xi$

After 100 iterations, obtain f_val=-0.3243 when Kw=0.3846, PDR=645.6097kw, T1=0.0562s, T3=0.0643.

5. Conclusion

The control strategy design about the coordination of demand response and UPFC to suppress low frequency oscillation

Finally, optimize related control parameters of the coordinated control system containing demand response and UPFC by using the genetic algorithm and obtain the optimal control parameter, being the optimal strategy for the coordination of demand response and UPFC to control low frequency oscillation in the power system .

References

[1] Black J W, Ilic M. Demand-based frequency control for distributed generation [C] // Power Engineering Society Summer Meeting. 2002:427-432.

[2] Lu N, Hammerstrom D J, Patrick S. PNNL-18998 Grid FriendlyTM Device Model Development and Simulation[R]. Washington: Pacific Northwest National Laboratory, 2009.

[3] Aggregated Modeling and Control of Air Conditioning Loads for Demand Response. Wei Zhang,Jianming Lian,Chin-Yao Chang,Kalsi, K. Power Systems . 2013

[4] Guo J, Crow M L, Sarangapani J. An Improved UPFC Control for Oscillation Damping[J]. IEEE Transactions on Power Systems, 2009, 24(1):288-296.

[5] Y.-Q. Bao, Y. Li, "FPGA-based design of grid friendly appliance controller," IEEE Trans. Smart Grid, vol. 5, no. 2, pp. 924-931, 2014.

[6] Liu Y W, Rau S H, Wu C J, et al. Improvement of Power Quality by using Advanced Reactive Power Compensation[J]. IEEE Transactions on Industry Applications, 2018, PP(99):1-1.

[7] Binwei J I, Chen X. Research on Reactive Power Compensation of Distribution Network Based on Particle Swarm Optimization Algorithm[J]. Power System & Clean Energy, 2016

[8] S. A. Pourmousavi, M. H. Nehrir, "Real-time central demand response for primary frequency regulation in microgrids," IEEE Trans. Smart Grid, vol. 3, no. 4, pp. 1988-1996, 2012.

IOP Conf. Series: Journal of Physics: Conf. Series **1061** (2018) 012023 doi:10.1088/1742-6596/1061/1/012023

Learning-Based Approach for Face Image Relighting

Ching-Ting Tu[1, *], Chin-Yu Chang[1] and Yi-Chung Chen[1]

[1]Department of Computer Science and Information Engineering, Tamkang University, Taiwan, R.O.C.

*cttu@mail.tku.edu.tw

Abstract. In this study, we propose a system for face image relighting. Given an unseen testing image, the system automatically changes its light condition. Such system is difficult since the shadow and lighting areas of a particular lighting condition are highly depended on the lighting source position and the facial 3D geometry. To solve this problem, we introduce a learning-based approach to construct the appearance correlation between two lighting conditions in order, which is then used to transform the lighting condition from the source lighting domain to the target lighting domain. First of all, the discriminative facial features of lighting and personal characteristics are selected. Consequently, these discriminative features are then used to guide the light reconstruction process. According to the experimental results, the proposed system, where the discriminative features are integrated in the synthesis process, makes the synthesized results preserve the personal characteristics of system input.

1. Introduction

We proposed a system for facial image relighting, i.e. changing the lighting condition of the input facial image to another specified lighting condition. However, the lighting condition of a facial image is highly depended on the position and direction of the lighting sources. Furthermore, even under the same light condition, shadow and highlight facial areas of images from different individuals are not equal. To solve such problem, we proposed a learning–based algorithm to transfer the lighting condition of the input image (the source condition) to a target condition. Such learning process requires a training dataset, where two facial images of the same individual taken under the source and target lighting conditions, respectively, are required. The proposed learning approach can apply to other testing samples not observed in the training dataset. The key issues are (i) learn the complex appearance correlation of images taken under two different lighting conditions. (ii) Personal characteristics of input individual, such as the shapes of nose or eye, should be preserved after changing the lighting condition.

1.1. Related work

There are several works for facial image relighting. For example, Shihet al.'s [1] proposed a lighting transformation system to relight facial image as a particular professional style. Such framework required a reference image first, and geometrically align any input testing image to the selected reference image. The lighting condition of input image is then mapped into the selected target lighting condition, e.g. the shadow and highlight areas of the reference image. However, the facial alignment process is generally not performed well for the shadow area, and the transformed result will be decreased. Work proposed in [2] is also required the face alignment process. The idea of this work is quite similar to work [1], it transfers the input lighting condition to that of the selected reference image. However, method [1] processed on the image lighting layer (the low frequency of facial image), on the hand, method [2] focused on the texture layer, which is the high frequency information containing the local detailed component. Approach in [3] calculated the lighting contrast between the input and reference images by the help of 3D facial model. In summary, these three methods are required a reference image. However, the personal geometry and the appearance characteristics of input and that of the selected reference may be totally different.

On the other hand, learning-based methods, such as [4] and [5], were proposed; method in [4] adopted kernel principal component analysis (kernel PCA) to learn the lighting correlation between two different lighting conditions from the training database. Method in [5] used Tensor PCA model to construct the subspace of different face individuals taken under various lighting conditions. Consequently, the constructed subspace is then used to predict the lighting of unseen input. However, according to the PCA property, the synthesized image of any unseen subject will be similar to appearances of training samples. In this study, we propose a learning-based approach for face image relighting, the aim is to preserve the personal characteristics of the input.

2. Learning-based synthesis system

A learning-based face relighting synthesis system is proposed in this study, where given a facial image with a particular lighting condition, a corresponding synthesized image under another lighting condition is produced. Such synthesis system is trained with the assistance of a training dataset containing pairwise lighting examples, where each one contains two images of the same individual taken under two particular lighting conditions. Especially, the proposed synthesis framework is cooperated with a feature extraction process, where facial regions that are invariant to cross-lighting conditions and significant for discriminating personal characteristics are extracted. The learned lighting-condition dependency feature map is then used to constrain the learning-based synthesis process, the proposed framework is illustrated in figure 1.

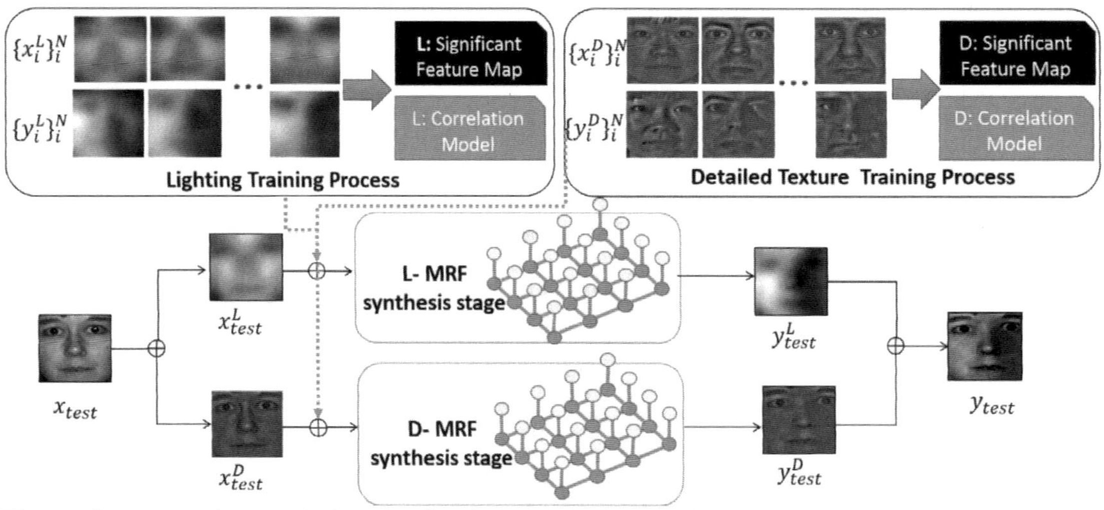

Figure 1. Proposed example-based relighting system. L: the lighting stage, D: the detailed texture stage, *x*: images taken under source lighting condition, and *y*: image taken under target lighting condition.

2.1. Pairwise dataset

Let the source (i.e. input) lighting condition as the style-*X* and the target (i.e. output) lighting condition as the style-*Y*. We collect *N* pairwise training samples from Yale B database [6], where each training individual contains two facial images taken under two specified lighting conditions *X* and *Y*, respectively. Denote the dataset $\boldsymbol{X} = [\boldsymbol{x}_1, \boldsymbol{x}_2, .., \boldsymbol{x}_N]$ as the *N* images of the style-*X*, and the dataset

$\boldsymbol{Y} = [\boldsymbol{y_1}, \boldsymbol{y_2}, \ldots, \boldsymbol{y_N}]$ as the images of style-Y. Each image (either $\boldsymbol{x_i}$ or $\boldsymbol{y_i}$) is decomposed into its lighting component ($\boldsymbol{x_i^L}/\boldsymbol{y_i^L}$) and the detailed texture component ($\boldsymbol{x_i^D}/\boldsymbol{y_i^D}$). For example, the decomposing process of image $\boldsymbol{x_i}$ is:

$$\boldsymbol{x_i^L} = \boldsymbol{x_i} \otimes \boldsymbol{G}(\boldsymbol{\sigma}) \text{ and } \boldsymbol{x_i^D} = \boldsymbol{x_i} - \boldsymbol{x_i^L} \quad (1)$$

Where $\boldsymbol{G}(\boldsymbol{\sigma})$ is the Gaussian blurring function with kernel size $\boldsymbol{\sigma}$; $\boldsymbol{y_i}$ can be decomposed as $\boldsymbol{y_i^L}$ and $\boldsymbol{y_i^D}$ in the similar way.

2.2. Significant feature extraction

Two training datasets, $\boldsymbol{L} = \{(\boldsymbol{x_i^L}, \boldsymbol{y_i^L})\}_{i=1}^{N}$, and $\boldsymbol{D} = \{(\boldsymbol{x_i^D}, \boldsymbol{y_i^D})\}_{i=1}^{N}$, are then used for extract the light and detailed texture features, respectively. For the light case, the feature selection process is to extract the significant facial patches that can classify the images taken under the lighting condition $\boldsymbol{X}$ (style-$\boldsymbol{X}$) or from the lighting condition $\boldsymbol{Y}$ (style-$\boldsymbol{Y}$). On the other hand, for the detailed texture case, the goal is to select patches that are lighting invariant for face identification in order to preserve the personal characteristics. The feature selections for lighting and detailed textures are processed by the Adaboost algorithm [7], respectively.

In the feature extraction process of the lighting case, the dataset $\boldsymbol{L}$ is used, where each images is divided into $\boldsymbol{N_k}$ overlapped patches. Accordingly, we adopted $\boldsymbol{N_k} * \boldsymbol{N}$ patches from the $\{\boldsymbol{y_i^L}\}_{i=1}^{N}$ as features for the feature pool of the Adaboost algorithm. In such classifier learning process (the feature selection process), $\{\boldsymbol{y_i^L}\}_{i=1}^{N}$ is used as positive samples and $\{\boldsymbol{x_i^L}\}_{i=1}^{N}$ as the negative one. Feature value of each sample (either positive $\boldsymbol{y_i^L}$ or negative $\boldsymbol{x_i^L}$) for a particular feature is defined as the cosine similarity between this feature patch and the corresponding location of this feature patch in this sample. figure 2(c) illustrates the selected feature maps. In this case, the negative samples (style-X) are taken under frontal source lighting condition, shown in figure 2(a); the positive samples (style-Y) are taken from the target left lighting condition as figure 2(b). We used the 40%-feature map for the following synthesis stage.

One the contrary, for the detailed texture case, dataset D is used, where the (x_i^D, y_i^D) pairwise sample from the same individual is denoted as the positive sample, on the other hand, (x_i^D, y_j^D) pairwise sample from two different individuals is denoted as the negative sample. In such case, the feature pool are collected from the $N_k * N$ patches in $\{y_i^D\}_{i=1}^{N}$. Feature value of a pairwise sample (x_i^D, y_j^D) for a particular feature is defined as the cosine similarity between this feature patch and the corresponding patches at the same location in these two detailed images, respectively, and the product of these two cosine similarities is denoted as the feature value. figure 2(d) illustrates the selected feature maps of detailed texture. We used the 60%-feature map for the following synthesis stage.

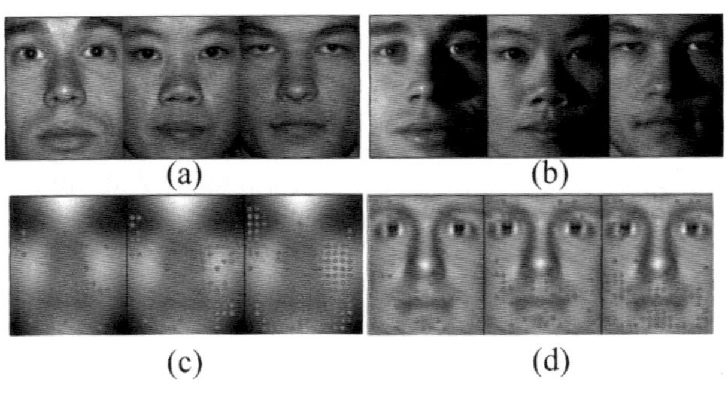

Figure 2. (a) example images of style X; (b) corresponding images of style Y; (c) lighting feature map selected by Adaboost algorithm; (d) detailed feature map. Note that, for (c) and (d), from left to right, are the top 20%, 40%, and 60% feature maps, respectively.

2.3. Synthesis stage

For any input image $\boldsymbol{x_{test}}$ that taken under the source lighting condition (style-$\boldsymbol{X}$) is decomposed as its lighting component ($\boldsymbol{x_{test}^L}$) and the detailed texture component ($\boldsymbol{x_{test}^D}$) by using (1) first. The goal of synthesis stage is to transfer (estimate) the lighting and detailed style of $\boldsymbol{x_{test}^L}$ and $\boldsymbol{x_{test}^D}$ to $\boldsymbol{y_{test}^L}$ and $\boldsymbol{y_{test}^D}$, respectively. The estimation process is based on the Markov random field (MRF) approach [8]. For the synthesis stage, the proposed MRF graph containing $\boldsymbol{N_k}$ observation nodes, which are corresponded to the $\boldsymbol{N_k}$ patches of each image, and each observation node has one corresponding hidden node, which represents the estimated patch appearance. In work [8], the candidate(s) of each hidden node are selected from the training dataset; however, in our approach, the candidates are either collected from the training dataset or is the same as the observation appearance. Functions of these two candidate types are different, where the former one makes the final result contain the target *Y*-style, while the later one makes the synthesized result preserve more personal characteristics of system input $\boldsymbol{x_{test}}$.

The objective function of the proposed synthesis module is modelled by the MRF formulation:

$$\boldsymbol{p}(\boldsymbol{x_{p1}}, \boldsymbol{x_{p2}}, \dots, \boldsymbol{x_{pN_k}}, \boldsymbol{y_{p1}}, \boldsymbol{y_{p2}}, \dots, \boldsymbol{y_{pN_k}}) \propto \prod_{(pi,pj)} \boldsymbol{\psi}(\boldsymbol{y_{pi}}, \boldsymbol{y_{pj}}) \prod_{pk} \boldsymbol{\phi}(\boldsymbol{x_{pk}}, \boldsymbol{y_{pk}}) \quad (2)$$

Where the joint probability function of the input, i.e. the observation layer $\boldsymbol{x} = (\boldsymbol{x_{p1}}, \boldsymbol{x_{p2}}, \dots, \boldsymbol{x_{pN_k}})$, and the final result, i.e. the hidden layer $\boldsymbol{y} = (\boldsymbol{y_{p1}}, \boldsymbol{y_{p2}}, \dots, \boldsymbol{y_{pN_k}})$, are modelled by using the product of data terms and the smoothness terms. The data term (or the observation probability), $\boldsymbol{\phi}(\boldsymbol{x_{pk}}, \boldsymbol{y_{pk}})$, is used to evaluate the probability of using $\boldsymbol{y_{pk}}$ as the state candidate of the *pk*'th hidden node given the *pk*'th observation node as $\boldsymbol{x_{pk}}$. The smoothness tem (or the transition probability), $\boldsymbol{\psi}(\boldsymbol{y_{pi}}, \boldsymbol{y_{pj}})$, is used for the probability of using $\boldsymbol{y_{pi}}$ as the state candidate of the *pi*'th hidden node given state candidate of its neighbour hidden node (the *pj*'th node) as $\boldsymbol{y_{pj}}$. In the proposed framework, the observation function of the *pk'th* hidden node and the *pk'th* observation node is defined as:

$$\boldsymbol{\phi}(\boldsymbol{x_{pk}}, \boldsymbol{y_{pk}}) = \boldsymbol{exp}\left(-\frac{\left\|x_{pk} - \theta_x(y_{pk})\right\|_2^2}{(\eta_L \sigma_k^L)^2}\right), \quad (3)$$

Where $\boldsymbol{\sigma_k^L}$ is determined by the cross verification; $\boldsymbol{\theta_x}(\boldsymbol{y_{pk}})$ is the corresponding training patch of style-*X* of $\boldsymbol{y_{pk}}$. The transition function of the *pi'th* and *pj'th* hidden nodes is defined as:

$$\boldsymbol{\psi}(\boldsymbol{y_{pi}}, \boldsymbol{y_{pj}}) = \mathbf{exp}\left(-\frac{od(y_{pi}, y_{pj})}{(\eta_L' \sigma_k^L)^2}\right), \quad (4)$$

Where such probability value is used for measuring the appearance difference between the overlapped regions of the *pi'th* patch and its neighbour (the *pj'th* node), i.e. $\boldsymbol{od}(\boldsymbol{y_{pi}}, \boldsymbol{y_{pj}})$. In short, the function of $\boldsymbol{\phi}(\ )$ is to select satiable candidates that with similar appearance as the observation data, while the function of $\boldsymbol{\psi}(\ )$ is used for confirming the smoothness (or the continuous appearance) between neighboring patches. Such formulation is applied for both the lighting and the detailed texture component synthesis, respectively.

In the optimization framework of the MRF method [8], the state candidates are selected from the training dataset. To be specified, using the lighting case as an example, the pairwise images of each example in the training dataset $\boldsymbol{L} = \{(\boldsymbol{x_i^L}, \boldsymbol{y_i^L})\}_{i=1}^{N}$ are decomposed as $\boldsymbol{N_k}$ patches, respectively. i.e. $\boldsymbol{x_i^L} = (\boldsymbol{x_{i,p1}^L}, \boldsymbol{x_{i,p2}^L}, \dots, \boldsymbol{x_{i,pN_k}^L})$ and $\boldsymbol{y_i^L} = (\boldsymbol{y_{i,p1}^L}, \boldsymbol{y_{i,p2}^L}, \dots, \boldsymbol{y_{i,pN_k}^L})$. Accordingly, for the *pi*'th hidden node, there are *N* candidates from $\{\boldsymbol{y_{i,pi}^L}\}_{i=1}^{N}$, where each candidate has one corresponding patch from

IOP Conf. Series: Journal of Physics: Conf. Series **1061** (2018) 012023 doi:10.1088/1742-6596/1061/1/012023

$\{\boldsymbol{x}_{i,pi}^{L}\}_{i=1}^{N}$, i.e. $\boldsymbol{y}_{i,pi} = \boldsymbol{\theta}_{x}(\boldsymbol{y}_{i,pi})$ in (3). These candidates are then selected for maximizing the objective function formulated in (2), (3), and (4), by using the belief propagation algorithm. However, if all the state candidates are from the training samples as the original method in [8], the insufficient texture variance in the training dataset may cause unexpected synthesized results (as shown in figure 3(c)).

To solve such problem, in the proposed method, some specified observation nodes of the input, e.g. $\boldsymbol{x}_{test,pi}^{L}$, are also used as the state candidates of the corresponding hidden nodes. To be specified, these nodes are determined by the significant features learned by Adaboost algorithm in sec 2.2.

That is, the patch locations that are corresponding to significant feature maps are used to determine the state candidates should be from the training dataset, e.g. $\{\boldsymbol{y}_{i,pi}^{L}\}_{i=1}^{N}$, or directly assigned as its observable patch, $\boldsymbol{x}_{test,pi}^{L}$. For example, for the lighting case, the location of *pk*'th patch that belongs to region of the selected features indicate that its candidates should be from the training dataset (i.e. $\{\boldsymbol{y}_{i,pk}^{L}\}_{k=1}^{N}$) for inferring the appearance that are included in the input; on the contrary, candidates of other hidden nodes are from their corresponding observation $\boldsymbol{x}_{test,pk}^{L}$. A similar process is also applied for texture synthesis; however, the candidate nodes that corresponding to the location of selected are directly assigned as their observations, e.g. $\boldsymbol{x}_{test,pi}^{D}$, as these regions are invariant to lighting for face identification. Candidates of other nodes are form the training dataset (i.e. $\{\boldsymbol{y}_{i,pk}^{D}\}_{k=1}^{N}$). Final synthesized results based on the selected feature map are shown in figure 3(d).

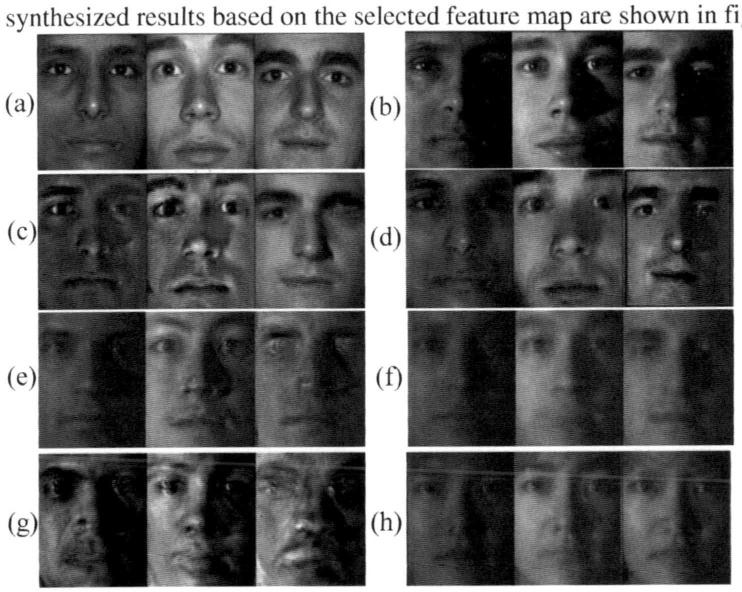

Figure 3. (a) thee testing examples of style *X*; (b) corresponding images of style *Y* (ground truth); results synthesized by (c) MRF model [4], (d) proposed method, (e) DCM model [9], (f) 2DDCM model [10], (g) LS method, and (h) CCA model [11].

3. Experimental results

We validated our method using the Yale B database [6]. There are 38 subjects in this dataset, where each subject has two pairwise image taken under two specified lighting conditions, i.e. the style *X* (as shown in figure 3(a)) and the style *Y* (as shown in figure 3(b)). In our setting, 32 subjects are selected for the training purpose, while the rest 6 while the rest 6 subjects are used for the testing. In figure 3, we also compare the proposed approach with other learning-based regression method for predicting facial structure under the style *Y*, i.e. direct combined model (DCM) [9] (figure 3(e)), 2DDCM [10] (figure 3(f)), least squares regression (LS) (figure 3(g)), and canonical correlation analysis (CCA) [11] (figure 3(h)). In figure 4, we also evaluated the proposed approach for 4 cases of pairwise lighting conditions, synthesized results of three example testing subjects are provided.

4. Conclusion

We proposed an example-based face relighting system. Using Adaboost approach, the discriminative features are selected to boost the performance of synthesized result. The synthesis process is based on the MRF approach, where different from the existing approaches, the proposed optimization framework is modified by cooperating with the discriminative feature selection process in order to preserve the personal characteristics of the system input. Our method outperformed existing approaches in the current dataset.

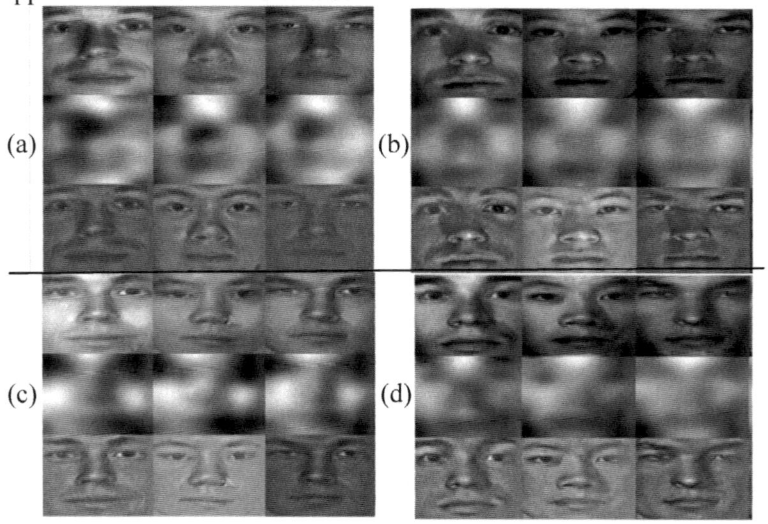

Figure 4. Synthesized results (style-*Y*) of testing examples, where training samples for (a) and (b) are from the same pairwise dataset, but style *X* and style *Y* are interchanged. Same for the results of (c) and (d). Note that, for each case, the first row: final synthesized results; the second row: synthesized lighting results; the third row: results of the detailed texture.

References

[1] Shih Y., Paris S, Barnes C, Freeman W F, and Durand F 2014 Style transfer for headshot portraits *ACM Trans. Graph.* pp. 1-148

[2] Chen X, Wu H, Jin X, and Zhao Q 2013 Face illumination manipulation using a single reference image by adaptive layer decomposition. *IEEE Transactions on Image Processing* pp. 4249-4259

[3] Wang Y, Zhang L, Liu Z, Hua G, Wen Z, Zhang Z, and Samaras D 2009 Face relighting from a single image under arbitrary unknown lighting conditions *IEEE Trans. Pattern Anal. Mach. Intell* pp. 1968-1984

[4] Xie X, Zheng W S, Lai J, Yuen P C, and Suen C Y 2011 Normalization of face illumination based on large-and small-scale features *IEEE Transactions on Image Processing* pp. 1807-1821

[5] Geng X, Smith-Miles K, Zhou Z H, and Wang L 2011 Face image modelling by multilinear subspace analysis with missing values *IEEE Transactions on System, Man, and Cybernetics, Part B (TSMC-B)* pp.881-892

[6] Georghiades A S, Belhumeur P N, and Kriegman D J 2001 From few to many: illumination cone models for face recognition under variable lighting and pose *IEEE Trans. Pattern Anal. Mach. Intell. 23(6)* pp. 643-660

[7] Viola P A and Jones M J 2004 Robust real-time face detection *International Journal of Computer Vision* pp. 137-154

[8] Liu C, Shum H Y, Freeman W T 2007 Face hallucination: theory and practice *International Journal of Computer Vision* pp. 115-134

[9] Tu C T and Lien J J 2010 Automatic location of facial feature points and synthesis of facial sketches using direct combined model *IEEE Trans. Systems, Man, and Cybernetics, Part B* pp. 1158-1169

[10] Tu C T, Ho M C, and Lin M Y 2017 A new approach for face hallucination based on a two-dimensional direct combined model *Pattern Recognition* pp. 1-20

[11] Yang W, Yi D, Lei Z, Sang J, Li S Z 2008 2D-3D face matching using CCA *FG* pp. 1-6